Chance in Biology

Chance in Biology

Using Probability to
Explore Nature

Mark Denny and
Steven Gaines

Princeton University Press, Princeton and Oxford

Copyright © 2000 by Princeton University Press
Published by Princeton University Press, 41 William Street, Princeton, New Jersey 08540
In the United Kingdom: Princeton University Press, 3 Market Place,
Woodstock, Oxfordshire OX20 1SY

Library of Congress Cataloging-in-Publication Data

Denny, Mark W., 1951–
Chance in biology : using probability to explore nature / Mark Denny and
Steven Gaines.
p. cm.
Includes bibliographical references.
ISBN 0-691-00521-4
1. Biomathematics. 2. Probabilities. I. Gaines, Steven, 1951– II. Title.

QH323.5 .D46 2000
570′.1′5192—dc21   00-036687

This book has been composed in Times Roman

The paper used in this publication meets the minimum requirements of
ANSI/NISO Z39.48-1992 (R1997) *(Permanence of Paper)*

www.pup.princeton.edu

Printed in the United States of America

10 9 8 7 6 5 4 3 2 1

For my parents,
who are teaching me to be optimistic
when playing life's odds.
— M.D.

For Peggy, Erin, and Andrea,
for unvarying support and love.
— S.G.

Contents

Preface

The basic idea for this book was hatched over a pitcher of beer at the 1992 Benthic Ecology Meeting in Newport, Rhode Island. After a day spent deciphering ANOVA tables as one ecologist after another grappled with the variability in his or her data, we were waterlogged with inferential statistics and in the mood to complain. It all seemed so pessimistic; scientists railing against the rising tide of natural fluctuation, everyone wishing that the world were just a bit more orderly. Wouldn't it be fun, we thought, to stop fighting the randomness of the real world and put all that variability to work? Maybe it was just the beer, but our commiseration eventually led to one of those cartoon moments in life when each of us could see a lightbulb suddenly illuminate over the other's head. Stochastic processes could be our friend! And we were off and running.

Over the time that it has taken to write this text, we have had a lot of fun developing the idea that chance events can be used to make strong predictions about biology. Not the least of the enjoyment has been derived from those social encounters at meetings and parties in which a student or colleague asks the bland question, "So, what are you working on these days?" For seven years now, we have had the rare privilege of answering, "Oh, I'm writing a statistics text." The amusement comes when we watch the mental and facial contortions as the student or colleague fumbles for a courteous reply. Of course, to most biologists the obvious retort is, "Why would anyone want to write a book on the driest subject known to man?" but few people have been blunt enough to actually say so. The ensuing silences have allowed us to practice our sales pitch, and after several years of honing, we present it to you here.

Why Should You Read This Book?

For decades physicists, mathematicians, engineers, and fluid dynamicists have used the intrinsic stochastic nature of the world to their benefit. Through the application of the theory of probability, they have been able to make important predictions about how things work, and these predictions, embodied in rules such as the ideal gas laws and Fick's equations for diffusion, are standard tools in many areas of science and technology. A few areas of biology have embraced this approach (neurophysiology and population genetics come to mind), but the vast majority of biologists cling to the notion that in the advancement of science, the stochastic variability of nature is a "bad thing." At the very least, it is annoying. How discouraging it must be to think that the random fluctuations of life serve primarily to obscure our knowledge of how life works. This intrinsically pessimistic approach is not made better by the standard biostatistics courses that are taught to undergraduates. These tend to be heavily laden with an endless list of statistical recipes, with scant attention paid to the theory of probability on which these recipes are based. No wonder most biologists think of statistics as desiccated and inscrutable.

To put this into the vernacular, we feel that it is time for biologists to stop being reactive in issues of natural variability; it is time to start being proactive. In this text, we hope to provide the reader with a sense of the good things that are possible when probability theory is applied to biology. If we are successful, you will come away with a glimpse of the joy to be had in bringing order out of chaos, the feeling of achievement in making an important biological prediction based on the intrinsic variability of nature.

What about All Those Equations?

These rewards cannot be had without some work on your part. There is no way around the fact that this is a mathematical textbook, and if you are like most biologists, the math might make you a bit uncomfortable. In structuring the text, we have tried to minimize this problem. First, this book is geared to an audience with a modest mathematical background. We assume that you have taken an introductory calculus course at some point in your career, but we do not assume that you use this knowledge on a day-to-day basis. In other words, we suppose that you know that the derivative of a function is equal to its slope, but we do not assume that you could readily calculate anything but the most basic derivatives. Similarly, we assume that you have a "feeling" that the integral of a function is equal to the area under the curve, but again we don't expect you to be able to calculate a complex integral.

Second, the mathematics in the book is gently progressive. We start off simply, then gradually move on to more challenging material. Wherever possible, we include intermediate steps in calculations and provide at least informal derivations so that you will know where the final, bottom-line equations come from. In the rare case where a derivation is just too complicated or technical to present here, we give reference to more specialized texts in which the full derivation can be found. If you forget what a symbol stands for, there is a separate index to direct you to the page on which that symbol is defined.

We hope that this approach will make this text accessible to a wide audience. Perhaps by working through the examples presented here you can knock some of the rust off your mental gears and rediscover how much fun a little math can be when applied to real-world problems. In case you care to test your understanding as you go along, we have included brief problem sets at the ends of chapters 2 through 8. Answers to the questions are provided in chapter 9, but you will be doing yourself a favor if you make an honest attempt to solve each problem first before consulting the cheat sheet.

This Book Is Not about Inferential Statistics

Finally, we feel compelled to warn you that this book is not a substitute for an introductory text on biostatistics. Hypothesis testing, error propagation, the analysis of variance, and correlation and regression analysis are all-important topics, but they are not covered here. We do provide an informal introduction to probability theory (a subject shared with the better texts on biostatistics), but then we take a different path, as outlined above. This is not to say that the approach taken here is at odds with that of inferential statistics. Indeed, we have for years used this text as an adjunct to courses in introductory statistics. The two viewpoints on stochastic processes are complementary, and we feel that you, the reader, can benefit from exposure to both.

Acknowledgments

We owe a debt to many people for their help in writing this text. Perhaps foremost, we are indebted to Howard Berg. His classic text *Random Walks in Biology* (1983) spurred our interest in the practical application of probability theory, and served as the stepping-off point for chapters 5 and 6. He also reviewed an early draft of the text, and the final version has benefited greatly from his insightful comments. Ron King appeared out of nowhere to introduce us to the statistics of extremes, and we again thank him for the fun that ensued. A host of students and colleagues have provided advice, references, or have read all or portions of the text; we thank them all: Dennis Baylor, Andrew Brooks, Brian Gaylord, William Gilly, Ben Hale, Elizabeth Nelson, Joanna Nelson, Mike O'Donnell, Leslie Osborne, Sunil Puria, Christian Reilly, Loretta Roberson, Allan Stewart-Oaten, Dale Stokes, and Stuart Thompson. These folks have been of immense help in our attempt to rid the manuscript of errors; any remaining goofs (and all the lame attempts at humor) are ours alone.

Loretta Roberson brought the sarcastic fringehead to our attention, and Brian and Briana Helmuth opened our eyes to the bipolar smuts. Dale Stokes performed the calculations and visual manipulations for figure 8.3, and Freya Sommer lent her artistic talents and broad biological knowledge to the drawings.

And finally, we thank our families for their forbearance: Sue, Katie, and Jim; Peggy, Erin, and Andrea.

Chance in Biology

1

The Nature of Chance

1.1 *Silk, Strength, and Statistics*

Spider silk is an amazing material. Pound for pound it is four times as strong as steel and can absorb three times as much energy as the KEVLAR from which bullet-proof vests are made (Gosline et al. 1986). Better yet, silk doesn't require a blast furnace or a chemical factory for its production; it begins as a viscous liquid produced by small glands in the abdomen of a spider, and is tempered into threads as the spider uses its legs to pull the secretion through small spigots. Silk threads, each a tenth the diameter of a human hair, are woven into webs that allow spiders to catch prey as large as hummingbirds. Incredible stuff!

The strength and resilience of spiders' silks have been known since antiquity. Indeed, anyone who has walked face first into a spiderweb while ambling down a woodland path has firsthand experience in the matter. Furthermore, the basic chemistry of silk has been known since early in this century: it is a protein, formed from the same amino acids that make our skin and muscles. But how does a biological material, produced at room temperature, get to be as strong as steel and as energy-absorbing as KEVLAR? Therein lies a mystery that has taken years to solve.

The first clues came in the 1950s with the application of X-ray crystallography to biological problems. The information provided by this technique, used so successfully by James Watson and Francis Crick to deduce the structure of DNA, allowed physical chemists to search for an orderly arrangement of the amino acids in silks—and order they found. Spider silk is what is known as a crystalline polymer. As with any protein, the amino-acid building blocks of silk are bound together in long chains. But in silks, portions of these chains are held in strict alignment—frozen parallel to each other to form crystals—and these long, thin crystals are themselves aligned with the axis of the thread. The arrangement is reminiscent of other biological crystalline polymers (such as cellulose), and in fact can account for silk's great strength.

It cannot, however, account for the KEVLAR-like ability of spider silk to absorb energy before it breaks. Energy absorption has two requirements: strength (how

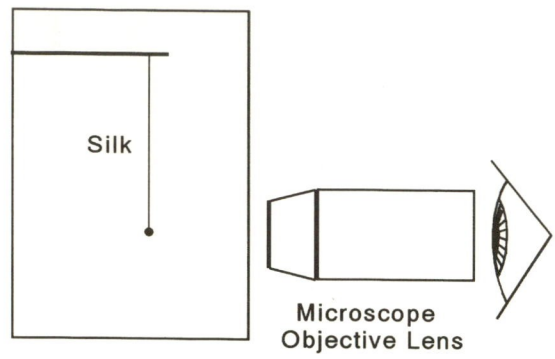

Fig. 1.1 A schematic diagram of the apparatus used to measure the change in length of a piece of spider silk. The vial in which the silk is suspended is itself immersed in a temperature-controlled water bath, and the microscope is used to track the location of the silk's free end

much force the material can resist) and extensibility (how far the material can stretch). Many crystalline polymers have the requisite strength; cellulose, for example, is almost as strong as silk. But the strength of crystalline polymers is usually gained at the loss of extensibility. Like a chain aligned with its load, a protein polymer in a crystal can extend only by stretching its links, and these do not have much give. Cellulose fibers typically can extend by only about 5% before they break. In contrast, spider silk can extend by as much as 30% (Denny 1980). As a result of this difference in extensibility, spider silk can absorb ten times more energy than cellulose. Again, how does silk do it?

This mystery went unsolved for 30 years. As powerful a tool as X-ray crystallography is, it only allows one to "see" the ordered (aligned) parts of a molecule, and the ordered parts of silk (the crystals) clearly don't allow for the requisite extension. If silk is to be extensible, some portion of its molecular structure must be sufficiently disordered to allow for rearrangement without stretching the bonds between amino acids in the protein chain. But how does one explore the structure of these amorphous molecules?

The answer arrived one dank and dreary night in Vancouver, British Columbia, as John Gosline performed an elegant, if somewhat bizarre, experiment (Gosline et al. 1984). He glued a short length of spider silk to a tiny glass rod, and, like a sinker on a fishing line, glued an even smaller piece of glass to the thread's loose end. Pole, line, and sinker were then placed in a small vial of water so that the spider silk was vertical, held taut by the weight at its end (fig. 1.1). The vial was stoppered and placed in a second container of water in front of a microscope. By circulating water through this second bath, Gosline could control the temperature of the silk without otherwise disturbing it, and by watching the weight through the microscope he could keep track of the thread's length. The stage was set.

Slowly Gosline raised the temperature. If silk behaved like a "normal" material, its length would increase as the temperature rose. Heat a piece of steel, for

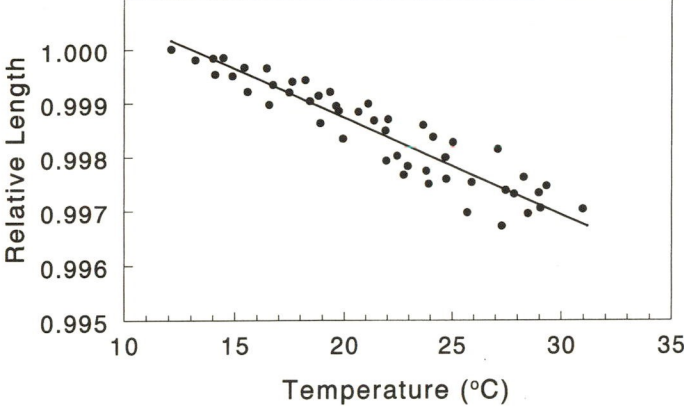

Fig. 1.2 The length of a piece of spider silk decreases slightly as temperature increases, an effect explained by the random rearrangement of the amorphous portions of the material. (Data from Gosline et al. 1984)

instance, and it will expand. In contrast, if the molecules in the amorphous portions of the silk are sufficiently free to rearrange, they should behave differently. In that case, as the temperature rose, the amorphous protein chains would be increasingly rattled by thermal agitation and should become increasingly contorted. As with a piece of string, the more contorted the molecules, the closer together their ends should be. In other words, if the amorphous proteins in silk were free to move around, the silk should get shorter as the temperature was raised. This strange effect could be predicted on the basis of statistics and thermodynamics (that's how Gosline knew to try the experiment) and had already been observed in man-made rubbery materials. Was spider silk built like steel or like rubber?

As the temperature slowly drifted up—10°C, 12°C, and higher—the silk slowly shortened (fig. 1.2). The amorphous portions of silk are a rubber! From this simple experiment, we now know a great deal about how the noncrystalline portions of the silk molecules are arranged (fig. 1.3), and we can indeed account for silk's great extensibility. Knowing the basis for both silk's strength and extensibility, we can in turn explain its amazing capacity to absorb energy—knowledge that can potentially help us to design man-made materials that are lighter, stronger, and tougher.

There are two morals to this story. First, it is not the orderly part of spiders' silk that makes it special. Instead, it is the molecular disorder, the random, ever-shifting, stochastic arrangement of amorphous protein chains that gives the material its unique properties. Second, it was a knowledge of probability and

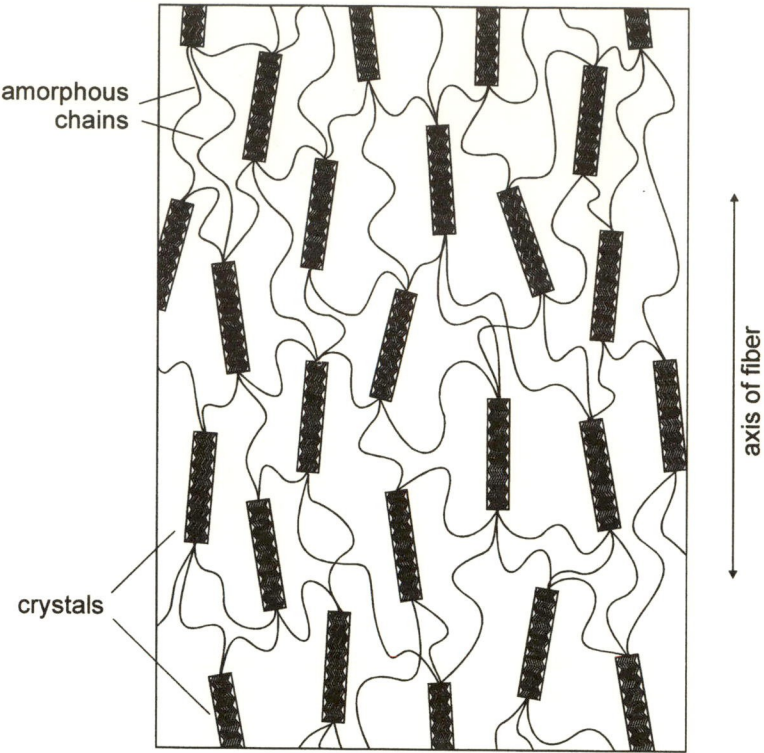

FIG. 1.3 The molecular architecture of spider silk. Crystals of ordered proteins are aligned with the silk fiber's axis, providing the material with great strength. Randomly arranged (amorphous) protein chains connect the crystals in an extensible network.

statistics that allowed Gosline to predict the consequences of this disorder and thereby perform the critical experiment. As we will see, the theory of probability can be an invaluable tool.

Probability theory was originally devised to predict the outcome in games of chance, but its utility has been extended far beyond games. Life itself is a chancy proposition, a fact apparent in our daily lives. Some days you are lucky—every stoplight turns green as you approach and you breeze in to work. Other days, just by chance, you are stopped by every light. The probability of rain coinciding with weddings, picnics, and parades is a standard worry. On a more profound level, many of the defining moments of our lives (when we are born, whom we marry, when we die) have elements of chance associated with them. However, as we have seen with spider silk, the role of chance in biology extends far beyond the random events that shape human existence. Chance is everywhere, and its role in life is the subject of this book.

1.2 *What Is Certain?*

As an instructive example, imagine yourself sitting with a friend beside a mountain stream, the afternoon sun shining through the trees overhead, the water babbling as it flows by. What can you say with absolute certainty about the scene in front of you? Well, yes, the light will get predictably dimmer as the afternoon progresses toward sunset, but if you look at one spot on the river bank you notice that there is substantial short-term variation in light intensity as well. As sunlight propagates through the foliage on its way to the ground, the random motion of leaves modulates the rays, and the intensity of light on the bank varies unpredictably both in space and in time. Yes, a leaf falling off a tree will accelerate downward due to the steady pull of gravity, but even if you knew exactly where the leaf started its fall, you would be hard-pressed to predict exactly where it would end up. Turbulent gusts of wind and the leaf's own tumbling will affect its trajectory. A close look into the stream reveals a pair of trout spawning, doing their instinctive best to reproduce. But even with the elaborate rituals and preparations of spawning, and even if all the eggs are properly fertilized, there is chance involved. Which of the parents' genes are incorporated into each gamete is a matter of chance, and which of the millions of sperm actually fertilize the hundreds of eggs is impossible to predict with precision.

Even the act of talking to a friend is fraught with chance when done next to a mountain stream. The babbling sound of the brook is pleasing because it is so unpredictable, but this lack of predictability can make communication difficult. Somehow your ears and your brain must extract from this background noise the information in speech.

So, chance in life is unavoidable. Given this fact, how should a biologist react? In many disciplines, the traditional reaction is to view the random variations of life as a necessary evil that can be exorcised (or at least tamed) through the application of clever ideas and (as a last resort) inferential statistics. Even then we are taught in our statistics classes to abhor unexplained variation. In a well-designed experiment, the less *chance* involved in the outcome, the better!

There is an alternative, however: the approach taken by Gosline in his experiment on spider silk. If chance is a given in life, why not use it to our advantage? In other words, if we know that a system will behave in a random fashion in the short term and at small scale (as with the random thermal motions of protein chains in silk), we can use this information to make accurate predictions as to how the system will behave in the long run and on a larger scale. Therein lies the thread of our tale. The diffusion of molecules, the drift of genes in a population, the longevity of phytoplankton, all include a large element of chance, and we will see why. How soft can a sound be before no animal can detect it? How fast must a mouse move in the moonlight before no owl can see it? We will be able to make predictions.

Before we embark on this exploration, we need to discuss briefly the nature of variation and which of nature's variations will (and will not) be included here.

1.3 *Determinism versus Chance*

One of Sir Isaac Newton's grand legacies is the idea that much about how the universe works can be precisely known. For example, if we know the exact mass of the moon and Earth and their current speed relative to each other, Newtonian mechanics and the law of gravitation should be able to tell us exactly where the moon is relative to Earth at any future time. As a practical matter, this is very close to being true. We can, for instance, predict solar and lunar eclipses with reasonable accuracy centuries in advance. Processes such as the moon's orbital mechanics are said to be *deterministic*, implying that, given sufficient knowledge of the initial state of a system, its future is determined exactly.

In fact, good examples of real-world deterministic processes are difficult to find. As our example of an afternoon spent observing a mountain stream is meant to convey, many of the processes that seem simple when described in the abstract (the variation in light intensity with the position of the sun, the downward acceleration of an object falling from a height) are exceedingly complex in reality. Details (rustling leaves and atmospheric turbulence) inevitably intrude, bringing with them an element of unpredictability. In some cases, the amount of variability associated with a process is sufficiently small that we are willing to view the system as being deterministic, and accept as fact predictions regarding its behavior. The physics of a pendulum clock, for instance, is so straightforward that we are content to use these machines as an accurate means of measuring time. In biology, few systems are so reliable, and deterministic behavior can be viewed at best as a polite fiction. As you might expect from the title of this book, deterministic processes will have no place here.

If a system or process is not deterministic, it is by definition *stochastic*. Even if we know *exactly* the state of a stochastic system at one time, we can never predict *exactly* what its state will be in the future. Some element of chance is involved. Unlike pregnancy and perfection, stochasticity can manifest itself to a variable degree. Many stochastic processes are approximately predictable with just a minor overlay of random behavior. The light intensity at our mountain stream is an example. Yes, there are minor random short-term fluctuations, but if we were to take 5-minute averages of the light level at the forest floor, they would closely follow predictions based on knowing the elevation of the sun. In other cases, the predictability of a system is negligible, and chance alone governs its behavior. The movement of molecules in a room-temperature gas is a good example. Both types of systems will be included in our exploration.

As a practical matter, the dividing line between "deterministic" and "stochastic" is open to interpretation. For example, it is common practice (both in sports and introductory texts on probability theory) to accept the flip of a coin as a chance proposition, a stochastic process. But if you know enough about the height above the ground at which the coin is flipped, the angular velocity initially imparted to the coin, and the effects of air resistance, it should be possible to decide in advance whether the coin will land heads up. Indeed, much of what we accept as stochastic may well be deterministic given sufficient understanding of the mechanism involved. In this respect, the line between "deterministic" and "stochastic" is often drawn as a matter of convenience. If the precise predictions that are possible in theory are too difficult to carry out in practice, we shift the line a bit and think of the process as being stochastic.

This is not to imply that *all* processes are deterministic, however. As far as physicists have been able to divine, there are aspects of nature, encountered at very small scales of time and space, that are unpredictable *even in theory*. For example, there are limits to the precision with which you can know both the velocity and the location of an object (this is a rough statement of Heisenberg's uncertainty principle). In other words, if you could know exactly where an electron is at some point in time, you couldn't know what its velocity is. Conversely, if you know exactly what its velocity is, you can't know its position. In either case, you can't predict exactly where the electron will be even a short time in the future. This is the strange realm of quantum mechanics, where chance reigns and human intuition is of little use. In this text we make scant use of the principles of quantum mechanics (a single, brief mention of the unpredictability of light emission in chapter 8). We introduce the subject here only to note that there is indeed a dividing line between "deterministic" and "stochastic," even if it is fuzzy.

1.4 *Chaos*

In recent years, this dividing line has become even fuzzier. Beginning in the late 1970s, a wide variety of physical systems that should behave deterministically were found in fact to behave unpredictably. These systems are said to exhibit *deterministic chaos*, or just *chaos* for short. But if they are deterministic, how can they be unpredictable? This apparent conflict is solved by the fact that chaotic systems are exquisitely sensitive to the state in which they are started.

Consider a "normal" deterministic system, something like the flight of a baseball through still air. In this case, if we know the initial speed of the ball (20 m s^{-1}), its initial location (home plate in Fenway Park, Boston), and its direction of motion (45° to the horizontal), we can predict where the ball will land (just beyond second base). If we make a small error in the measurement of any of these initial conditions (say, the ball is moving at 20.001 rather than 20.000 m s^{-1}), the error in predicting the ball's landing is concomitantly small.

If the motion of a baseball were chaotic, however, its flight would be quite different. Every time a chaotic ball were launched at exactly 20 m s^{-1} and an angle of exactly 45° from the center of home plate, it would land in the same spot near second base; the system is deterministic. If, however, a chaotic ball were launched with even a slight error in any of these initial conditions, its eventual landing spot would be drastically different. A shift from 20.000 to 20.001 m s^{-1} might cause it to land in the town square of, say, Emporia, Kansas. Granted, every time the ball is launched at *exactly* 20.001 m s^{-1}, it ends up in the same place in Kansas, so the system is still deterministic, but it is *extremely* sensitive to the initial conditions.[1]

This hypothetical example is intended only to provide an intuitive "feel" for the character of a chaotic system; it is sorely lacking as a definition. Unfortunately, a compact, formal definition of chaos is not easy to come by, and a lengthy explanation would be out of place here. If you want to pursue this intriguing field further, we suggest that you consult Moon (1992).

The strong sensitivity to initial conditions described above can make real-world chaotic systems appear to be stochastic. Shifts in initial conditions that are too small to be measured can cause the behavior of the system to fluctuate drastically, and these fluctuations can reasonably be assigned to "chance." The primary disadvantage of treating a chaotic system as if it were stochastic is a loss of insight. Once a process is stamped with the title "random," it is easy to stop looking for a mechanistic cause for its behavior. At present, however, there are few alternatives for understanding many chaotic systems, and in this text processes that actually may be chaotic will be treated as if they were stochastic.

---

[1] It is interesting to note that the motion of the planets, which has long been cited as the classical example of deterministic mechanics, is in fact chaotic. Because each planet is subject to a gravitational pull from all other planets, there is the possibility that at some time in the future the alignment of the solar system may be such that one of the planets could be thrown substantially off its present orbit, and this potential makes it virtually impossible to predict accurately where the planets will be at a given date in the future. See Peterson (1993) for an enlightening and readable discussion of chaos in planetary physics.

1.5 *A Road Map*

Our exploration of chance in biology begins in chapter 2 with a brief review of the theory of probability, culminating in a discussion of Bayes' formula and the difficulty of testing the general population for a rare disease. In chapters 3 and 4 we move from the examination of single events to the probabilities associated with large numbers of events. To this end, we introduce the notion of a Bernoulli trial and the binomial, geometric, and normal distributions.

With these basic tools in hand, we then examine the multifaceted role of random walks in biology (chapters 5 and 6). We explore the mechanics of molecular diffusion and use them to predict the maximum size of plankton in both water and air. An analogy between molecular diffusion and genetic drift allows us to predict the time expected before an allele becomes fixed or lost in a population. The nature of three-dimensional random walks is examined, leading to an explanation of why your arteries are elastic, how horses hold their heads up, and how scallops swim. It is here that we find out exactly why heated spider silk shrinks.

We then shift gears in chapter 7 and expand on our knowledge of probability distributions by exploring the statistics of extremes. We show why it is possible (but unlikely) that you will have your eardrums burst at a cocktail party, and how to predict the size of waves crashing on a rocky shore. It is here, too, that we predict the absolute limits to human longevity, the likelihood of the next 0.400 hitter in baseball, and why jet engines only occasionally flame out.

And finally, in chapter 8 we explore how our ability to see and hear the world around us is affected by the inevitable thermal and quantum noise of the environment. The trade-off between spatial and temporal resolution in sight is explained, accounting for why it is so hard to catch a ball at dusk. We predict the lower limit to size in nerve cells, and show how random noise can actually allow nerves to respond to signals that would otherwise be undetectable.

2

Rules of Disorder

The theory of probability is a formal branch of mathematics with elegant theorems, complicated proofs, and its own book of jargon. Despite these potential obstacles, people use probability informally nearly every day. When we play games, decide what to wear by glancing at the morning sky, or pick the route we will take to get across town during rush hour, we often rely on crude perceptions of probability to make decisions in the face of uncertainty. Even the most math-phobic individuals occasionally use elementary aspects of probability theory to guide their actions. Unfortunately, such primitive applications of probability are often misguided and can lead to illogical decisions. Our intuition is not a viable substitute for the more formal theory of probability.

Examples of this are evident when we play games of chance. People differ substantially in skill, leading some players to win (and others to lose) far more frequently than they should by chance alone. Most often, what we think of as skill in such games is simply a measure of how accurately a player's actions are consistent with an understanding of probability theory—unless, of course, their success relies on cheating. By analogy, when chance events play an important role in the design, function, or behavior of organisms, our skill in interpreting patterns in nature depends on our understanding of the theory of probability.

In this chapter, we briefly develop a set of definitions, rules, and techniques that provides a theoretical framework for thinking about chance events. For the sake of brevity, we focus on the aspects of probability theory that are most critical to the issues raised in the rest of this book. For more in-depth coverage of probability we recommend Feller (1960), Isaac (1995), or Ross (1997).

2.1 *Events, Experiments, and Outcomes*

Every field of science and mathematics has its own vocabulary, and probability is no exception. Unfortunately, probability theory has the added feature of assigning technical definitions to words we commonly use to mean other things in everyday life. To avoid confusion, it is crucial that we speak the same language, and, to that end, some definitions are necessary.

FIG. 2.1 The sarcastic fringehead. The upper panel shows the fish in repose. In the lower panel, two fringeheads engage in a ritual bout of mouth wrestling

In probability theory, the focus of our attention is an *event*. A more formal definition is given below, but put simply, an event is something that happens with some degree of uncertainty. Typically, books on probability theory use as examples events such as getting a one on the roll of a die, flipping a coin five times in a row without getting any heads, or sharing the same birthday with someone else at a party. These types of events are useful because they represent activities you can easily duplicate or imagine. As we have suggested, however, the uses of probability theory are far broader than playing games or matching birth dates; a large number of environmental and biological issues critically depend on the occurrence of uncertain events. Let's start with two biological examples where chance plays an important role.

2.1.1 SARCASTIC FISH

One of the most ferocious fish found along the Pacific Coast of North America is the sarcastic fringehead (fig. 2.1). Although it rarely exceeds a foot in length, the fringehead has an enormous mouth, a pugnacious temperament, and the wary respect of fishermen, who have been known to do "amusing little dances while 6 in. long fish clamp sharp teeth around their thumbs" (Love 1991).

Fringeheads typically live in holes or crevices in the rocky substratum. They aggressively defend these shelters by lunging at anything that approaches, snapping open their capacious mouths. When the intruder is another fringehead looking for a new shelter, the two individuals often enter into a ritual match of "mouth wrestling" with their sharp teeth interlocked (Stokes 1998). As with many ritualized fights in animals, these matches are a relatively benign mechanism for establishing dominance, and the larger of the two individuals inevitably wins the battle and takes over the shelter. But sarcastic fringeheads seem to be poor judges of size. Due perhaps to poor eyesight, an inflated perception of their own bulk, or both, the fish appear incapable of accurately evaluating the size of another individual until they begin to wrestle.

Now suppose you are a fringehead guarding your shelter. Along comes another fringehead and that old, instinctive urge to dominate rises up within you. You lunge out and commence to wrestle. With your mouths pressed together it is quickly clear that you are substantially larger than your opponent (just as you thought!), and the intruder scurries away. Your shelter is safe. Later, a second fringehead arrives. Again, you rush to defend your shelter, but this time your luck runs out. You aren't quite the fish you thought you were, your mouth is smaller than his, and you end up homeless.

The stage is now set for a few basic definitions. In the vocabulary of probability theory, these wrestling matches are *experiments*.[1] Experiments are simply processes that produce some observation or measurement. Since you cannot predict the result of the wrestling experiments with complete certainty before you leave your shelter, we call these wrestling matches *random experiments*. Every time you repeat the experiment of defending your home, there is a single *outcome* (that is, one of the several possible results). The set of all possible outcomes for an experiment is called the *sample space* for that experiment. In the case of fringehead wrestling, there are only two possible *elementary outcomes*, success (*s*) or failure (*f*), and these together form the sample space.

Let's now turn our attention to another example of chance in the interaction among organisms.

2.1.2 BIPOLAR SMUT

If you asked the average person on the street what he knew about smut and sex, he would probably feign ignorance and scurry away in search of a

[1] Note that the use of the word "experiment" in probability does not imply hypothesis testing as it might in inferential statistics or most fields of science.

policeman. If by chance you happened to ask a mycologist, however, you would evoke a *very* different response. We found this out when we naively inquired of a friend of ours if she knew anything interesting about reproduction in lower plants, and received in return an energetic lecture on the wonders of sex in the smuts.

Smuts, it turns out, are parasitic fungi in the order Ustilaginales. They commonly infect vascular plants, including a variety of economically important grains, and as a result have been studied in depth. Reproduction in smuts is bizarre by vertebrate standards. To be precise, smuts do not have separate sexes, but they nonetheless reproduce sexually.

This poses some potential problems. One of the advantages of separate sexes is that gametes from one individual cannot fuse with another gamete from the same individual. This eliminates the most extreme form of *inbreeding*—mating with yourself. In the smuts, individuals produce haploid spores,[2] each of which fuses with another spore to create a new generation of smut. But instead of having male and female individuals that produce gametes of distinctly different sizes (as you find in most animals and plants), smuts typically produce spores that are morphologically indistinguishable from one another.

Lacking discrete sexes, how do smuts avoid inbreeding? As with many other fungi, the smuts promote mating with other individuals (*outcrossing*) through the use of *compatibility genes*. In a simple case, a single gene locus has two or more alleles that determine whether spores are compatible for fusing. If the allele present at the compatibility locus differs between two spores, they can fuse; if the alleles are the same, they cannot.

For example, let's identify the different alleles at the compatibility locus by letters (e.g., *a*, *b*, *c*, etc.). Because spores are haploid, each has a single compatibility allele. If one spore has allele *a* and another spore has allele *b* (or *c*, or *d*, or anything but *a*), they can fuse. Smuts with this mating system are termed *bipolar* because two alleles determine mating compatibility. Other fungi (the tetrapolar fungi) take their sex to an even higher level by having a mating system with two separate compatibility loci. Here, spores must differ at both loci before they can fuse.

Note that in the bipolar mating system all adult smuts must be heterozygous at the compatibility locus. The only way they could be homozygous is if both of the spores that fused to form the adult had the same allele, and if they had the same allele, they could not fuse.

---

[2] In sexually reproducing organisms, each adult has two sets of chromosomes, one from each parent, and the organisms are therefore in a *diploid* state. In the process of manufacturing gametes by meiosis, the number of chromosomes is cut in half, and these special reproductive cells (spores in this case) are in a *haploid* state.

Now, imagine two smuts on wheat (or rye). Smut 1 has compatibility alleles *a* and *b*. The other (smut 2) has compatibility alleles *c* and *d*. You could not find a more perfectly matched couple! Let's use these fertile fungi to perform an experiment.

It's mating time, and each of the smuts produces a multitude of spores, which you carefully collect in a bag. After mixing the spores thoroughly to randomize which spores are in contact, you empty the bag onto a microscope slide. There are thousands of pairs of contiguous spores, and as you watch, some fuse successfully and begin to grow. Others remain unfused, never to know the life of a smut. What can we predict as to which spores will fuse and which will not?

We begin by noting that each pair of spores can be viewed as another example of a random experiment. The outcomes of the experiment are successful fusion or the failure to fuse, and success and failure depend on chance. But in this case, we have additional information that can be applied to the problem. We already know that the results of these fusion experiments depend on the underlying genetics. Therefore, let's examine the outcomes in terms of the genotypes of the spores at the compatibility locus.

For any given spore in the experiment described above, there are four possible results corresponding to the four compatibility alleles found in the parents (*a*, *b*, *c*, and *d*). Since there are four possible results for each spore in the pair, there is a combined total of sixteen possible outcomes (see box 2.1).

Box 2.1. Why sixteen and not eight?

If there are four possibilities for the first spore and another four for the second spore, why do we multiply instead of add to get the total number of possibilities? The answer can be seen by stepping through the problem. Assume the first spore has allele *a*. How many possibilities are there for the second spore? The answer is four, since the second spore could have any of the four alleles. Similarly, if the first spore has allele *b*, there are still four possible values for the second spore. Each possibility for spore 1 has four possibilities for spore 2. Therefore, there are four sets of four, which is sixteen total outcomes. Notice that some of the sixteen possible outcomes are functional duplicates because the order of the alleles doesn't matter to the genotype of the offspring (e.g., *a* fusing with *b* and *b* fusing with *a*). Sometimes, order does matter, and we will deal with the consequences of duplicate outcomes a little later.

Furthermore, the outcomes of this smut fusion experiment are more complex than those in our fringehead wrestling experiment. Recall that when fringeheads wrestle, the outcome is *elementary* (also known as simple or indecomposable):

either success or failure. In contrast, each genetic outcome of the smut experiment has two parts, each part corresponding to the allele from a potential parent. If we view the results of randomly picking a single spore as yielding an elementary outcome (= the spore's allele), the outcome of a fusion experiment is *complex*; it includes two elementary outcomes.

Complex outcomes can be represented by an ordered set. In this case, our outcome is an ordered pair of elementary outcomes (x, y), where x is the allele of one spore and y is the allele of the other spore. This fusion experiment is analogous to grabbing two socks blindly from a drawer. Spores with different incompatibility alleles are like socks with different colors. Using this sock analogy, successful sex in smuts is like picking two socks from the drawer that do not match.

Using these definitions and examples, we can now define an event more formally. An *event* is a *set of outcomes* from an experiment that satisfies some specified criterion. We use these two terms (event, set of outcomes) interchangeably. The most basic events associated with an experiment are those that correspond to a single outcome[3]—for example, winning a mouth wrestle, which we call event *Win*; or getting a pair of spores, both with allele a, which we call event AA. Each of these basic events includes only one of the following possible outcomes:

$$Win = (s)$$

$$AA = (a, a).$$

Events can also include several outcomes. For example, suppose you do not care which alleles a pair of spores has as long as they match. Given the mating strategy of smuts, the event *Match* could also be described as "a sexually incompatible pair of spores." The set associated with *Match* includes four of the possible sixteen outcomes of our experiment:

$$Match = \big\{(a, a), (b, b), (c, c), (d, d)\big\}. \tag{2.1}$$

In this fashion, we can define a wide variety of events related to the same fusion experiment.

2.1.3 DISCRETE VERSUS CONTINUOUS

So far, the sample spaces we have discussed include a small number of possible outcomes. These are examples of *discrete sample spaces*. Discrete sample

---

[3] To help keep things as clear as possible, we'll use the convention of writing the names of *events* beginning with a capital letter and the names of *outcomes* in all lower-case letters.

spaces have a *countable* number of outcomes, by which we mean that we can assign an integer to each possible outcome. Any sample space with a finite number of outcomes is discrete, but some discrete sample spaces include an infinite number of outcomes. For example, we could ask how many wrestling matches a fringehead will enter before it loses for the first time. If we assume the fringehead is immortal and that there is an inexhaustible supply of challengers coming by, it is possible (although highly unlikely) that the fringehead could continue winning forever. The sample space for possible outcomes in this experiment is thus infinite, but countable.

Other experiments have an infinite number of possible outcomes that are *uncountable*. This commonly occurs in experiments where, by necessity, each outcome is measured using real numbers rather than integers. For example, the time it takes a predator to capture its prey can be measured to the fraction of a second, the average annual rainfall in Cincinnati can be calculated to a minute portion of an inch, and a compass heading can be determined to a minuscule part of a radian. In such situations, there are theoretically an infinite number of possible outcomes within any measurement interval, no matter how small the interval. Such experiments produce a *continuous sample space* with an uncountably infinite number of possible outcomes. We will return to continuous sample spaces in chapter 4.

2.1.4 DRAWING PICTURES

To analyze chance events, it is often useful to view the sample space (the set of all possible outcomes) in a diagram. Traditionally, the entire sample space is represented by a box. In a discrete sample space, each possible outcome is then represented by a point within the box. For our sarcastic fringehead wrestling experiment, for instance, there would be only two dots in the box (one for a success, *s*, and one for a failure, *f*; fig. 2.2A). For our experiment in smut reproduction, the box has sixteen dots corresponding to the sixteen ordered pairs of alleles (fig. 2.2B).

Once the possible outcomes are drawn in the box, events can be represented as "disks," where each disk is a closed curve that includes the set of points, if any, that satisfy the criteria of the event. This type of diagram is called a *Venn diagram*, after its originator John Venn.[4] A Venn diagram for our smut experiment is shown in figure 2.2B. Here, the disk labeled *Match* depicts the event of a pair of spores that have matched compatibility alleles.

---

[4] *Venn in doubt, draw a diagram.* Upon graduating from college, John Venn became a priest for five years, after which he returned to Cambridge University as a lecturer in Moral Science. Venn later grew tired of logic and devoted his time to writing history books and designing new machines. His most intriguing invention was a device to bowl cricket balls. The machine was so good it clean bowled one of the top stars of the Australian cricket team four times.

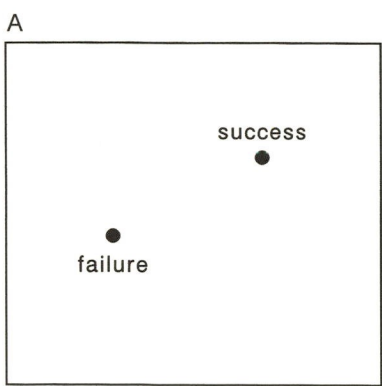
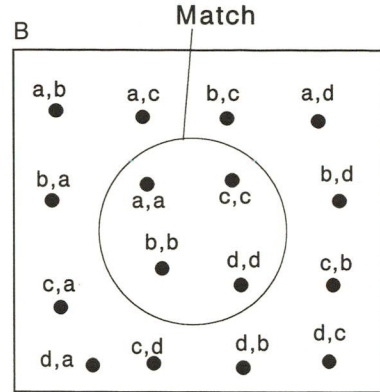

FIG. 2.2 Examples of Venn diagrams. Panel A depicts the complete sample space for a fringehead wrestling match. In this simple experiment, there are only two possible outcomes. Panel B shows the sixteen possible outcomes of the mating between smut 1 and 2. The event labeled *"Match"* includes the outcomes in which compatibility alleles are the same.

2.2 *Probability*

So far, we have thrown out a slew of definitions and drawn some potentially useful diagrams, but we have yet to touch on the central concept of our discussion, that of probability. In other words, we have characterized the possible outcomes of an experiment, but we currently have no means of estimating or predicting how frequently different outcomes might occur. In this section, we will develop the notion of the probability of an event.

For a particular manifestation of a random experiment, each outcome in the sample space has some possibility of occurring, but only one outcome can actually occur. The *probability* of a specific outcome is defined as the fraction of a large number of experiments that will yield this particular outcome.

DEFINITION 1 *Provided the number of experiments is very large:*

$$P(x) = \frac{number\ of\ occurrences\ of\ outcome\ x}{total\ number\ of\ repeated\ random\ experiments}.$$

This probability (denoted $P(x)$ for outcome x) must lie between 0 (the outcome never occurs) and 1 (the outcome always occurs).

In practice, we never know the precise probability of any uncertain event, but there are two general ways by which we can estimate its value: we can make an empirical estimate using repeated random experiments, or we can make a theoretical estimate using an idealized model of the random process.

Our experiment with smut sex provides an example. If the spores of smut 1 (with compatability alleles *a* and *b*) are mixed with spores of smut 2 (alleles *c* and *d*) and we repeatedly draw out pairs of spores at random, we find that each pairing occurs with equal frequency. Thus, each of the sixteen outcomes shown in figure 2.2B has a probability of 1/16.

A philosophical note is in order here. Given our definition, it is meaningless to talk about the probability of an experiment that cannot be repeated. If you carry out an experiment, an outcome results. But unless you can *repeat* the experiment, the number of occurrences of that outcome (= 1) must equal the number of trials (= 1). Thus, by our definition, the probability of an unrepeated experiment is exactly 1, in this context an uninformative number.

In contrast, as the number of repeated experiments grows, the frequency of occurrence becomes a better and better estimate of the actual probability of a given outcome for any single experiment. Formally, this rule is called the *Law of Large Numbers*. Fortunately, it expresses how most people intuitively think of the probability of a chance event.

2.3 *Rules and Tools*

Although estimating probabilities through the use of repeated random experiments is a common tool, there are many situations where this approach may be inaccurate, unacceptably expensive, unethical, or even impossible. For example, the number of experiments needed to get an accurate estimate of the probability may be inconveniently large, especially if you are trying to estimate the probability of rare events. In other cases, experimentation may be impractical since the event you are interested in may be something you are actively trying to avoid (e.g., an oil spill). You would not want to cause such events just to estimate their probability of occurrence. Finally, some questions may require experiments that society deems unethical. Examples include human trials of new drugs or surgical procedures where the expected risks are potentially large. To handle these cases, we need to develop models of probabilistic events. These models will be simplified abstractions of the real world, but they may help us to evaluate stochastic phenomena that we cannot study experimentally. Let's begin by considering how the probability of events builds on the probability of outcomes from an experiment.

2.3.1 EVENTS ARE THE SUM OF THEIR PARTS

Recall that an event is a set of outcomes that meets some criteria. If any outcome in the set occurs, the event occurs. As a result, the probability that an event will occur can be derived from the sum of the probabilities of all outcomes in its set:

Rule 1 $$P(\text{event } A) = \sum P(\text{outcomes in } A).$$

(Translation: If each of several different outcomes satisfies the criterion for an event, as in the event of getting a pair of smut spores with matching compatibility alleles, the probability of the event is simply the sum of the probabilities of the individual outcomes.)

To give some tangibility to this rule, let's return to the Venn diagram for our experiment in smut reproduction (fig. 2.2B). Here each of the points in the disk labeled *Match* satisfies the criterion for the event in which a pair of spores has matching alleles. Getting two *a* alleles *or* two *b* alleles *or* two *c* alleles *or* two *d* alleles are all satisfactory. Since each of these is a distinct outcome, we can simply add their respective probabilities (1/16) to obtain the overall probability of getting matching alleles: $4 \times 1/16 = 1/4$. In other words, the probability of the compound event *Match* is the sum of the probabilities of the individual outcomes enclosed by the disk in the Venn diagram. In fact, the probability of *any* event (no matter how complicated it may be) can be estimated if you can (1) identify the individual outcomes in the event and (2) if you know the probability of each of these outcomes.

Unfortunately, it is not always easy to both identify the individual outcomes in an event and know the probability of each outcome. As a result, we need to develop tools that allow us to estimate probabilities of sets we cannot measure directly.

Before leaving the subject of additive probabilities, we consider one corollary of rule 1:

Rule 2 $$\sum_{\text{all } i} P(x_i) = 1.$$

(Translation: The sum of the probabilities of all outcomes from an experiment equals one.)

This feature should be intuitive. As long as our sample space includes all possible outcomes, their probabilities must sum to one. In each experiment, *something* will happen. As we will see, rule 2 is used extensively for calculating the probability of complex events.

2.3.2 THE UNION OF SETS

As we have seen in the case of matching alleles, complex events can be formed from combinations of individual *outcomes*. Similarly, we can build even more complex events by combining *events*. Let's explore two examples that will be useful later.

Recall that when a pair of smut spores share the same compatibility allele, fusion cannot occur. As a result, the action of compatibility genes is to reduce inbreeding. Reduce, yes, but it does not totally preclude the fusion between two spores produced by the same parent. Since all smuts are heterozygous at the compatibility locus, all smuts produce two spore genotypes, and two spores of different types can successfully fuse even though they have the same parent. Thus, inbreeding. Now, suppose that you are interested in the event of getting an inbred offspring from one or the other of the parents in our smut mating experiment. How would you describe this event?

To answer this question, we focus on simpler (although still complex) events. There are two parents in our experiment, smut 1 (with compatibility alleles a and b) and smut 2 (with c and d). Let's let I_1 denote the event of getting an inbred offspring from smut 1. There are two outcomes in this event [(a, b) and (b, a)]. Similarly, I_2 is the event of getting an inbred offspring from smut 2 [(c, d), (d, c)], and our overall event (let's call it I for inbred offspring in general) includes all of the outcomes in *either I_1 or I_2*. I is therefore a combination of two simpler sets of outcomes. We call this combination a *union*. The union of two sets is typically denoted by the union operator, $\cup$. Therefore, $I = I_1 \cup I_2$. This union is shown as a Venn diagram in figure 2.3.

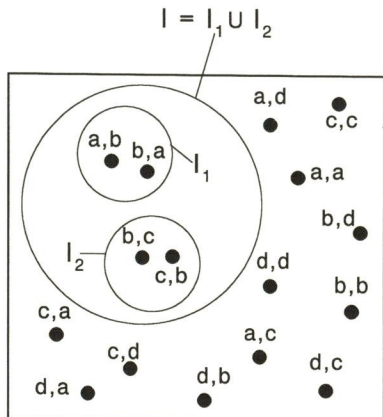

FIG. 2.3 A Venn diagram of the sample space for the smut sex experiment. The events labeled "I_1" and "I_2" include the outcomes corresponding to the inbred offspring of smuts 1 and 2, respectively. The event labeled "I" is the union of these two events. Note that events I_1 and I_2 do not share any outcomes.

For reasons that will become clear in a moment, a second example of the union of sets will be useful. Consider S_1, the event of getting an offspring of smut 1. S_1 is different from I_1 because in this case we do not care about the progeny's genotype. As long as one of its parents is smut 1, an offspring

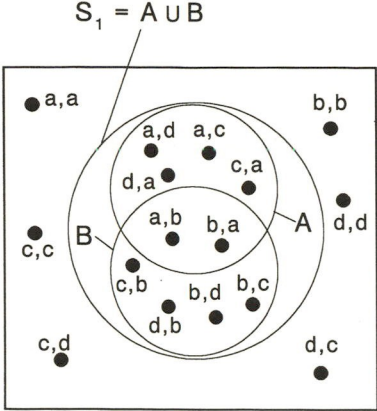

FIG. 2.4 A Venn diagram of the event S_1, the union between event A (outcomes containing allele a) and event B (outcomes containing allele b). Note that two outcomes are shared between events A and B.

qualifies. One simple way to describe this event follows from the realization that offspring of smut 1 must have either allele a or allele b (or both). Since smut 2 had neither of these two alleles, any new smut in our experiment with allele a (event A) and/or b (event B) must be an offspring of smut 1. Therefore, we can generate the event S_1 as the union between two events: $S_1 = A \cup B$. If you examine the Venn diagram in figure 2.4, you can see that six outcomes are found in each of these events. (Remember, aa and bb fusions do not produce offspring.) Unlike the previous example, these component events A and B do not have completely distinct outcomes. Their disks overlap because they share two offspring genotypes, (a, b) and (b, a). This overlap will have important consequences, which we discuss below.

2.3.3 THE PROBABILITY OF A UNION

Now, our interest in this exercise is to estimate the probability of the complex events described by the union of simpler events. Consider first $P(I)$, the probability of obtaining an inbred smut. Recall from rule 1 that the probability of an event is the sum of the probabilities of its individual outcomes. By analogy, perhaps we can use the sum of the probabilities of the two events, I_1 and I_2, to estimate the probability of I. Can we really add the probabilities of events the way we can add the probabilities of individual outcomes?

Our logic in formulating rule 1 was that because the outcomes in an event are distinct, the probability that the event occurs is the sum of the probabilities of the individual outcomes. If you examine the events I_1 and I_2 (fig. 2.3), you will find that they indeed do not overlap. Therefore, by the same logic as in rule 1, we can use the sum of the probabilities of I_1 ($= 1/8$) and I_2 ($= 1/8$) to calculate the probability of I:

$$P(I) = P(I_1) + P(I_2) = \frac{1}{8} + \frac{1}{8} = \frac{1}{4}. \tag{2.2}$$

So far, so good, but what happens when the events in a union *do* overlap? Here we can use our second example to see if the simple summation of probabilities still holds. As we noted above, the event of getting an offspring from smut 1 is the union of events A and B. But these events overlap (fig. 2.4). If we simply summed the probabilities of A and B, two outcomes in S_1 [$(a, b), (b, a)$] would get counted twice, once in A and once in B. By counting these outcomes twice, we would be overestimating the number of outcomes that qualify for event S_1. When this inflated number is inserted into our definition of probability, we would as a result overestimate the probability of S_1. Therefore, to calculate accurately the probability of S_1, we need to account for how shared outcomes affect the probability of the union of two or more sets. To do this, we use the concept of the intersection.

2.3.4 Probability and the Intersection of Sets

Outcomes that are shared by two sets are called the *intersection* between the sets. We denote the intersection by the operator, ∩. The intersection between the sets A and B includes two outcomes (see fig. 2.5):

$$A \cap B = \{(a, b), (b, a)\}.$$

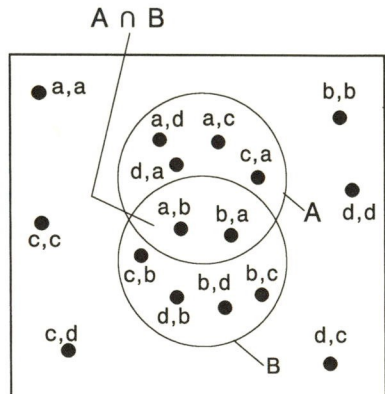

Fig. 2.5 A Venn diagram showing the intersection between event A (outcomes containing allele a) and event B (outcomes containing allele b). The interaction contains only those outcomes shared between events A and B.

More to the point, the intersection includes the same two outcomes that get counted twice if we add the probabilities of events A and B to arrive at the probability of S_1. The probability of S_1 should clearly include the probability of each of these two outcomes, but the probability of these shared outcomes should be included only once. As a result, we can estimate $P(S_1)$ from the sum of $P(A)$ and $P(B)$ by adjusting for the double counting of all outcomes shared by A and B (that is, by subtracting $P(A \cap B)$). Thus,

Rule 3   if $S_1 = A \cup B$, then $P(S_1) = P(A) + P(B) - P(A \cap B)$.

(Translation: If two events share outcomes, then the sum of the probabilities of the two events always exceeds the probability of the union of those two events. The difference is the probability of the shared outcomes, which erroneously gets counted twice.)

Note that this rule applies equally well to our initial example of inbred smuts. In this case, there is no overlap between I_1 and I_2, $P(I_1 \cap I_2) = 0$, and $P(I) = P(I_1) + P(I_2)$ as advertised.

2.3.5 THE COMPLEMENT OF A SET

Our laboratory experiments with reproduction in smuts greatly simplifies the real-world phenomenon by focusing on only two individuals. In an actual field setting, spores from one adult smut could potentially contact spores from a large number of other individuals. Furthermore, we have assumed here that there are only four alleles at the compatibility locus, but in actual populations the number of alleles may exceed a hundred. Both of these factors (more potential parents and more alleles) make it far more complicated to estimate probabilities in real populations.

For example, suppose you were a smut trying to estimate the probability that a particular individual spore you have produced could fuse with other spores encountered in the field. Let's assume that within the local smuts there are a hundred alleles at the compatibility locus, only one of which is contained in this particular spore. One approach to estimating your chance of producing offspring would be to sample the relative frequency of each of the other ninety-nine alleles in the population. Each of these alleles is compatible with the individual spore in question, and if you knew the probability of encounter for each of these genotypes, you could estimate the overall probability that this individual spore will successfully fuse. This approach is indeed possible, but it is *very* laborious.

A far simpler approach would be to focus on the single allele with which your spores could *not* fuse. Since the sum of the probabilities of all outcomes of an experiment must equal 1 (see rule 2), we can estimate the probability of an event by going in the back door, so to speak. If you could estimate the probability

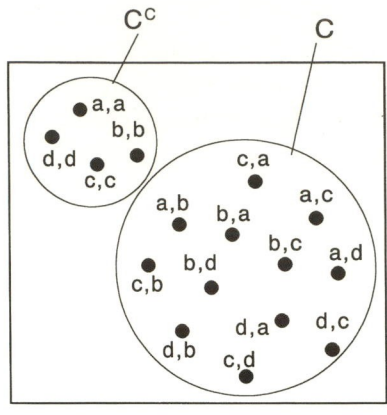

F_IG. 2.6 A Venn diagram showing the relationship between event C and its complement, C^c. All outcomes not in an event are in its complement.

of incompatibility, you could subtract it from 1 to estimate the probability of fusion. By turning the question on its head, a far simpler answer emerges.

This process leads us to define yet one more term. In probability theory, the *complement* of X is all outcomes in the sample space that are not in the set X. The complement is symbolized as X^c. In the example just discussed, we are interested primarily in the set C, the alleles with which our individual spore is compatible. But we carry out our calculation using C^c (the complement of C), the alleles with which our spore is *in*compatible (see the Venn diagram in fig. 2.6). The classic example of an event that is much easier to address as a complement is the question of estimating the probability that at least two individuals in a crowd share the same birthday. This probability is difficult to calculate directly, but quite simple to estimate using the complementary event. In other words, it is far easier to address the problem by estimating the probability that no two individuals share a birthday than it is to estimate the probability that at least two do share a birthday. We will leave the proof of this assertion as an exercise for you (see question 6 at the end of the chapter).

We note for future use the following fact:[5]

Rule 4 $$P(X) + P(X^c) = 1.$$

This follows from rule 2. Because all events are either in X or X^c, the sum of $P(X)$ and $P(X^c)$ must be 1.

---

[5] A fact well known to country music fans. As Clay Walker laments to his departed sweetheart, "The only time I ever miss you, honey/ is when I'm alone and when I'm with somebody."

2.3.6 ADDITIONAL INFORMATION AND CONDITIONAL PROBABILITIES

Let's now return to the event of getting an inbred offspring from smut 1 (I_1). Remember that this event includes two outcomes: $I_1 = \{(a, b), (b, a)\}$. Suppose there is one spore type whose genotype you can accurately identify. For example, suppose there is a rare mutation that, if present, causes spores with compatibility allele a to have a different color. As you scan a group of spores, you can thus identify with certainty the genotype of an occasional spore.

Suppose you randomly pick a pair of spores from those produced by smuts 1 and 2. You notice that one of the two spores has the color mutation. As a result, you know this spore has allele a. How does this additional information affect our estimate of the probability that this pair of fused spores will produce an inbred offspring of smut 1? It is clear that $P(I_1)$ must change based upon the additional information we now have, because only seven of the sixteen outcomes in our sample space are now possible— (a, a), (a, b), (b, a), (a, c), (c, a), (a, d), (d, a). At least one of the two alleles in each of these fusions is a.

We call the probability of an event based on the known occurrence of a separate event a *conditional probability*. It is denoted $P(X \mid Y)$, which is read "the probability of event X given that event Y occurs." Note that the conditional probability $P(X \mid Y)$ does not require that event Y happen first. For example, our task here is to find $P(I_1 \mid Color)$, the probability that a random pair of spores produces an inbred offspring of parent 1 given that (because of its color) at least one of the spores is known to have allele a. The number of inbred offspring produced (and therefore the process by which we calculate the probability of their production) is the same whether we observe the color of spores after or before they fuse. As long as we know for sure that event Y occurs, its temporal relationship to event X is irrelevant.

Let's examine a Venn diagram for our smut experiment (fig. 2.7) to see if we can figure out how the probability of I_1 will change given that we know that *Color* occurs. As we have seen, the event *Color* includes only seven outcomes from our original sample space of sixteen. In other words, if we see a spore with the color mutation, there is no chance for any of the nine outcomes that are not in *Color* to occur. In essence, the knowledge that the event *Color* has occurred shrinks our sample space.

We can use this information to calculate the modified probability of I_1 by using our existing techniques applied to this modified sample space. Our new, reduced sample space has only seven outcomes, and the event I_1 includes two of them [$(a, b), (b, a)$]. Thus, by our definition of probability, $P = 2/7$. In the long run it will be useful to express this conclusion in a more general fashion. According to rule 1, we should be able to estimate the probability of the event

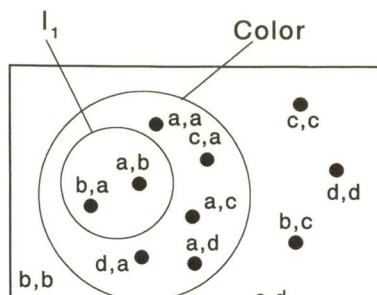

FIG. 2.7 An example of conditional probability. The event "*Color*" includes all outcomes that contain allele *a*. If we know that this event occurs (that is, we can see that a spore chosen at random is a different color than usual), this information affects the probability that event I_1 (an inbred offspring from smut 1) occurs.

I_1 by summing the probabilities of the two conditioned outcomes in I_1. This is indeed true, but we have a problem if our probabilities for the outcomes are based on our original sample space. In this case, each outcome has $P = 1/16$, and if we summed the probabilities of the seven possible outcomes in *Color*, they would not equal 1. In fact, they would sum to $P(Color)$. This suggests a solution. If we divide each outcome's probability by the total probability in our new sample space, $P(Color)$, the probabilities in the new sample space sum to 1.

Now all we need is an expression that defines those outcomes in our new sample space *Color* that also satisfy the event I_1, and our task will be complete. Solving this problem is easy, since the set of outcomes that satisfies both I_1 and *Color* is, by definition, the intersection of the two events. Thus, if we divide the probability of this intersection by $P(Color)$ to adjust the probabilities, we arrive at a formula for the probability of producing an inbred offspring of smut 1 given that we know one spore has allele *a*:

$$P(I_1 \mid Color) = \frac{P(I_1 \cap Color)}{P(Color)}$$

$$= \frac{2/16}{7/16} = \frac{2}{7}. \tag{2.3}$$

Or, in the general terms of any two events X and Y,

Rule 5
$$P(X \mid Y) = \frac{P(X \cap Y)}{P(Y)}.$$

(Translation: The probability that an event occurs, given that a second event occurs, is simply the probability that both events occur [that is, the probability of their intersection] divided by the probability of the event known to have occurred.)

One bonus from our effort with conditional probabilities is that we also gain a new formula to calculate the probability of the intersection between two events. If you rearrange the equation in rule 5, you get

$$P(X \cap Y) = P(X \mid Y) \cdot P(Y). \tag{2.4}$$

This formula provides an intuitively pleasing definition for the probability of the intersection of two events. Event Y occurs with probability $P(Y)$. Given that Y occurs, event X occurs with probability $P(X \mid Y)$. Therefore, the probability that both X and Y occur (that is, $P(X \cap Y)$) is the product of these two terms.

Now, in arriving at this conclusion, we have arbitrarily assumed that event Y is known to occur, but we could just as easily have assumed that X occurred. Thus, we can alternatively obtain the probability of the intersection of the two events using the probability that X occurs and the probability of Y given X. Therefore, it must also be true that

$$P(X \cap Y) = P(Y \mid X) \cdot P(X). \tag{2.5}$$

This equivalent form provides an important step to another useful rule in probability.

2.3.7 BAYES' FORMULA

Normally when we deal with equations, we try to simplify them as much as possible. This usually makes it easier to interpret what they mean. Sometimes, however, we can learn something by rearranging the equation into a more complex form. Let's return to our definition of a conditional probability (rule 5) for an important example. The probability that an event X occurs given that a second event Y occurs is equal to

$$P(X \mid Y) = \frac{P(X \cap Y)}{P(Y)}. \tag{2.6}$$

Let's expand this simple formula and see where it gets us. As we just discovered, the probability of the intersection between two events (the numerator here) can be written as $P(X \cap Y) = P(Y \mid X) \cdot P(X)$. To expand the denominator, we can use a trick based on the fact that the combination of an event (e.g., X) and its complement (X^c) includes all possible outcomes (rule 4). Therefore, we can write

$$P(Y) = P(Y \cap X) + P(Y \cap X^c). \tag{2.7}$$

In other words, all outcomes in Y must either be in X or its complement. If we further expand this formula for $P(Y)$ using eq. (2.4), we obtain

$$P(Y) = P(Y \mid X) \cdot P(X) + P(Y \mid X^c) \cdot P(X^c). \tag{2.8}$$

Now we are ready to thoroughly complicate rule 5. Substituting eq. (2.5) for the numerator and eq. (2.8) for the denominator into the formula for a conditional probability in eq. (2.6), we obtain

$$P(X \mid Y) = \frac{P(Y \mid X) \cdot P(X)}{P(Y \mid X) \cdot P(X) + P(Y \mid X^c) \cdot P(X^c)}. \tag{2.9}$$

This result may not seem like much of an accomplishment given the simple formula with which we started, but in fact this formula proves to be a very powerful tool.

This equation was originally proposed by Thomas Bayes, another English theologian and part-time mathematician. It is known as *Bayes' formula*. Notice that the equality shown here contains on its left side the probability for X conditioned on the occurrence of Y. In contrast, on the right-hand side, the conditional probabilities are all for Y conditioned on the occurrence of either X or X^c. In other words, the probability of one event conditioned on a second can be used to calculate the probability of the second event conditioned on the first. Therein lies the utility of Bayes' formula.

2.3.8 AIDS and Bayes' Formula

Your head is probably spinning from all these conditions, so let's consider an example to show how useful Bayes' formula can be. Isaac (1995) provides an excellent analysis of issues related to testing for HIV that shows how useful Bayes' formula can be.

For several years, blood and saliva tests have existed that can very accurately assess whether an individual has been infected with the AIDS virus, HIV. Although these tests are quite accurate, they occasionally make mistakes. There are two types of mistakes: false positives (where the individual tests positive but has never been exposed to the AIDS virus) and false negatives (where infection has occurred, but the test does not detect it). Experimental estimates of the likelihood of these events suggest that the existing tests for HIV are extremely accurate. If an individual is infected with HIV ($=$ event Inf), existing tests will be positive ($=$ event Pos) roughly 99.5% of the time. In other words,

$$P(Pos \mid Inf) = 0.995. \tag{2.10}$$

From this conditional probability, we can immediately calculate the probability of one type of mistake, a false negative. If a positive blood test for an infected individual occurs 99.5% of the time, this means that 0.5% of blood tests from infected individuals are negative (= event Neg). In other words,

$$P(FalseNegative) = P(Neg \mid Inf) = 1 - P(Pos \mid Inf) = 0.005. \qquad (2.11)$$

Using a similar procedure, we can estimate the probability of a false positive. In this case, we start with $P(Neg \mid NInf)$, the probability that we get a negative test result from a person who is not infected. In practice, estimating $P(Neg \mid NInf)$ is difficult because we need individuals who we know with certainty have not been infected (which requires a separate, unequivocal means of testing for HIV). Reasonable estimates of $P(Neg \mid NInf)$ using control groups with no likely risk of exposure to HIV suggest that this probability is roughly 0.995. As a result, the probability of a false positive is

$$P(FalsePositive) = P(Pos \mid NInf) = 1 - P(Neg \mid NInf) = 0.005. \qquad (2.12)$$

Therefore, the probabilities of test errors (either positive or negative) are extremely small for an individual test.

Now suppose that a misguided law is passed requiring all individuals to take a blood test for HIV infection, the intent being to quarantine infected individuals. If we select a random individual whose test is positive, what is the probability that this random individual is actually infected with HIV? If we translate this question into a conditional probability, we are asking what is

$$P(Inf \mid Pos)?$$

Notice that this conditional probability differs fundamentally from the conditional probabilities in our estimates of false positives and false negatives. Here we are trying to estimate the probability of actual infection conditioned on a test result. In eqs. (2.11) and (2.12), the reverse is true—we estimated the probability of a test result conditioned on a state of infection. This is a perfect opportunity to use Bayes' formula, which allows us to use probabilities conditioned on one event to estimate probabilities conditioned on another.

To simplify the interpretation, let's insert the events of this problem into Bayes' formula, eq. (2.9):

$$P(Inf \mid Pos) = \frac{P(Pos \mid Inf) \cdot P(Inf)}{P(Pos \mid Inf) \cdot P(Inf) + P(Pos \mid NInf) \cdot P(NInf)}. \qquad (2.13)$$

We have already estimated the conditional probabilities on the right side. All we need are estimates for $P(Inf)$, the fraction of the population that is infected.

The Centers for Disease Control estimated that there were 293,433 individuals reporting infection by HIV or AIDS in the United States in 1996. Given a population of approximately 270 million, this yields a rough estimate of $P(Inf) = 0.001$. Although this number surely underestimates the total number of infected individuals, it gives us a ballpark estimate of the probability of infection:

$$P(Inf) = 0.001, \quad \text{therefore } P(NInf) = 0.999. \tag{2.14}$$

Substituting these values into eq. (2.13), we obtain

$$P(Inf \mid Pos) = \frac{(0.995)(0.001)}{(0.995)(0.001) + (0.005)0.999} = 0.16. \tag{2.15}$$

This is an unexpected and disturbing result. Despite the fact that the blood test has only a minuscule chance of false positives (0.5%), a positive blood test implies only a 16% chance that an individual is actually infected. How can this be? Looking at Bayes' formula provides a clear explanation for this seeming paradox. Although individual tests have a low chance of error, most individuals who are tested are not infected with HIV. Therefore, we are multiplying a small probability of false positives by a large number of uninfected individuals. Even a minute probability of false positives for individual tests can in this circumstance produce many more false positives than true positives. As long as the disease is rare, even a very accurate test of infection will not be able to accurately identify infected individuals in a random test.

2.3.9 THE INDEPENDENCE OF SETS

In deriving Bayes' formula, we made repeated use of the information provided by conditional probabilities. That is, knowing that Y occurs gives us new insight into the probability that X occurs. There are cases, however, where the conditional probability of event X given event Y is the same as the unconditional probability of X. In other words, the added information of knowing that event Y occurs tells us nothing about the probability of event X occurring. Thus, if $P(X \mid Y) = P(X)$, the two events X and Y are said to be *independent* events. Independence of events turns out to be a very useful feature. Consider rule 5. If we use our definition of independence, we can substitute for the conditional probability on the left-hand side of the equation, $P(X \mid Y)$, to get

$$P(X) = \frac{P(X \cap Y)}{P(Y)}, \quad \text{if } X \text{ and } Y \text{ are independent.} \tag{2.16}$$

If we then multiply both sides of this equation by $P(Y)$, we obtain the famous *product rule* of independent events:

Rule 6   If X and Y are independent events, $P(X \cap Y) = P(X) \cdot P(Y)$.

(Translation: If two events do not influence each other's probability of occurring, the probability that both events will occur is simply the product of the probabilities that they individually occur.)

It turns out that this product rule can be extended to any number of independent events. If there are n independent events, the probability that all n events occur (i.e., the intersection of the n events) is the product of the n probabilities of the individual events.

As an example of the utility of the product rule, let's return once again to our sarcastic fringehead experiment. In this particular example, our fringehead won his first match and lost the second. But suppose we are interested in a more general question, the probability that the fringehead wins (and thereby retains his shelter) for the first time on the ith experiment. Let event $L_i =$ (retains shelter on match i). There are only two outcomes for each individual bout, success and failure, s and f. To keep track of the particular bout in which an outcome occurred, we use subscripts. Thus, s_{15} denotes a success in the fifteenth wrestling match.

Let's suppose that the probability of winning a match is p, and of losing is $q = 1 - p$. Assume also that the probability of winning is not affected by what happens during a previous wrestling match (i.e., bouts are independent—fringeheads do not become better wrestlers with more practice). Unless the fringehead retains its shelter in the first wrestling match, each L_i will be a series of $(i - 1)$ f's followed by a single s. For instance, if our fish first wins in the fifth match,

$$L_5 = (f_1, f_2, f_3, f_4, s_5). \tag{2.17}$$

Therefore, L_i is the same as

$$f_1 \cap f_2 \cap \cdots f_{i-1} \cap s_i. \tag{2.18}$$

Since the events associated with each match are independent, we can calculate the probability of this intersection using the product rule. The probability of s is p, and the probability of f is q. Therefore,

$$P(L_i) = q \cdot q \cdot q \cdot \ldots \cdot p \text{ where there are } (i - 1)q\text{'s.} \tag{2.19}$$

Thus,

$$P(L_i) = q^{i-1}p. \tag{2.20}$$

For example, if $p = 0.25$, the probability that our fish will first win on match i is shown in figure 2.8. There is only about an 8% chance that the first win will be in the fifth bout.

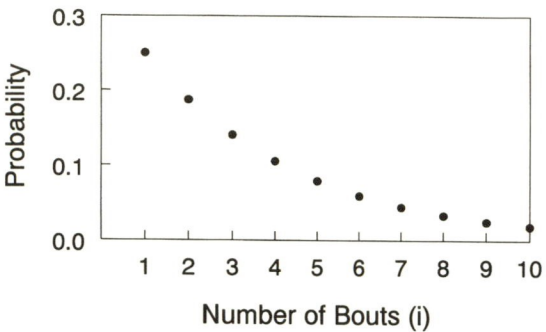

FIG. 2.8 The probability that a fringehead will first win on its ith match. Here we have assumed that in each bout the fish has a 25% chance of winning. This figure is a graphical representation of eq. (2.20).

2.4 *Probability Distributions*

In the last section we explored a variety of methods to calculate the probability for a particular outcome, and in the last example (how many bouts a fringehead will lose before winning) we even managed to derive an equation that describes the probability for each possible outcome. In other words, with a bit of diligent bookkeeping we can keep track of the frequency of occurrence of each and every outcome in the entire sample space, thereby associating each outcome with a probability. We can then lump outcomes into events and appropriately calculate the probability of each event.

This brings us to an important juncture in our exploration. Imagine writing down all the possible events in an experiment in one column of a table, and generating a companion column with each event's corresponding probability. This ensemble of paired outcomes and probabilities is called the *probability distribution* of the experiment.

Probability distributions will be of central importance throughout the rest of this book, and it will be best to take some time here to make sure that the concept is abundantly clear. Consider the same two smuts we have dealt with before, one with compatibility alleles a and b, the other with alleles c and d. Each produces an abundance of spores, and the spores are mixed randomly, and a single pair is chosen. What is the probability distribution for this reproductive experiment?

TABLE 2.1 The Probability Distribution for Spores Produced by the Mating of Two Smuts

Outcome	Probability	Outcome	Probability
a, a	1/16	c, a	1/16
a, b	1/16	c, b	1/16
a, c	1/16	c, c	1/16
a, d	1/16	c, d	1/16
b, a	1/16	d, a	1/16
b, b	1/16	d, b	1/16
b, c	1/16	d, c	1/16
b, d	1/16	d, d	1/16

Note: One smut has compatibility alleles *a* and *b*, the other has alleles *c* and *d*.

TABLE 2.2 The Probability Distribution for the Spores of the Two Smuts of Table 2.1

Event	Probability	Event	Probability
a, a	1/16	b, c	1/8
a, b	1/8	b, d	1/8
a, c	1/8	c, c	1/16
a, d	1/8	c, d	1/8
b, b	1/16	d, d	1/16

Note: In this case, the order of alleles is *not* taken into account.

First, we list the sample space for the simplest outcomes and their associated probabilities in table 2.1. In this case, each outcome has equal probability. But, as we have noted before, several of these simple outcomes are functionally equivalent [(a, b) and (b, a), for instance]. Thus, if we define an event as having a distinct allelic type *independent of order*, we have the list in table 2.2. The probability of an event in which alleles match is half that of events in which alleles are different.

We could simplify matters even more by again redefining what we mean by an event in the pairing of spores. Suppose that instead of tabulating the genotypes of paired spores we keep track of whether they fuse or not. In this case, the two possible events are *Fusion* and *Nonfusion*, and the corresponding probability distribution is as follows:

Event	Probability
Fusion	0.75
Nonfusion	0.25

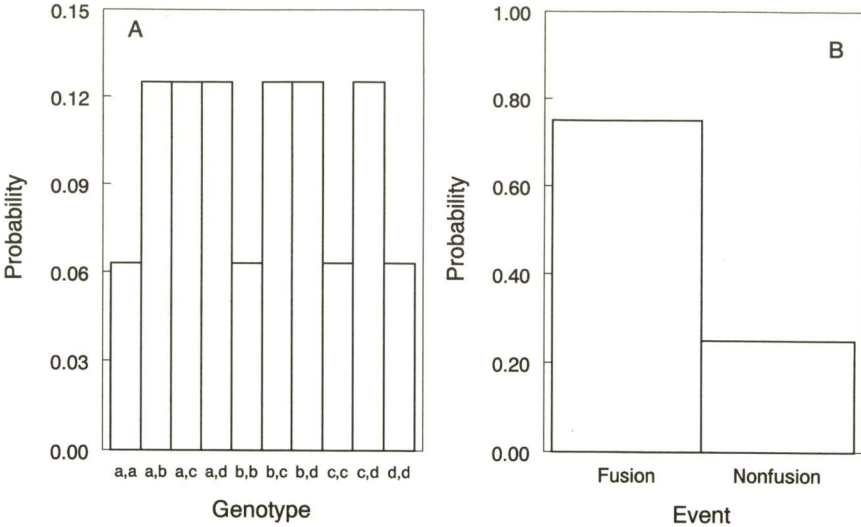

FIG. 2.9 Probability distributions for the mating between smuts 1 and 2. In panel A, events are defined in terms of the genotype of the paired spores. In panel B, events are defined with regard solely to whether the paired spores fuse. The same experiment can lead to different probability distributions depending on how events are defined.

Twelve of the sixteen possible ordered pairs are capable of fusion, so (given random pairing) fusion occurs three quarters of the time. The remaining pairs (those with matching compatibility alleles) do not fuse.

Tables of events and probabilities can be cumbersome when dealing with all but the smallest sample spaces. As a practical alternative, it is often handy to plot probability distributions graphically. This is traditionally done as a histogram. For example, the probability distribution of unordered allelic types is shown in figure 2.9A, and that for fusion/nonfusion in figure 2.9B. Alternatively, the probability distribution can be graphed as a scatter plot, as suggested by figure 2.8.

These simple probability distributions may seem obvious, and you are perhaps wondering why we are belaboring their existence. We have taken extra care in presenting the concept of probability distributions because these distributions are so fundamentally important. This importance lies in the fact that the probability distribution contains *all the information that can be known* about a given random experiment. Once you have established the criteria that define an event, have listed the sample space of possible outcomes, and have associated each event with a probability, you have in hand all the information that it is possible to obtain about the stochastic process in question.

This is not to say that this information cannot be processed further. For example, we will see in a moment how the probability distribution of simple events can be used to calculate the probability distribution of more complex events. In essence, this is what we did in working from the distribution of ordered allele pairs to the simple distribution of spore fusions. Knowledge of the probability distribution can also be used to calculate useful indices such as the average value of an experiment's outcomes and the typical variability of these outcomes. These uses of probability distributions will be covered in detail in chapters 3 and 4.

Note that it isn't necessary to present a probability distribution as a table. In many cases, a distribution can be described in more compact form by an equation. For example, eq. (2.20) describes the probability of event L (a fringe-head winning) occurring first on trial i. Because this equation is in essence a shorthand notation for a list of L, $P(L)$, it describes the probability distribution for L.

2.5 *Summary*

In this chapter we have developed a set of tools to help deal with estimating the probability of relatively complex events. We can use Venn diagrams to answer complex problems with a somewhat brute-force, graphical approach: draw a diagram with all possible outcomes and then choose those outcomes that are included in your event. Alternatively, by decomposing the problem into a set of simpler events, we can break what seems like a difficult question into a series of manageable tasks. Our arsenal of tools now allows us to examine the probability of (1) outcomes that are shared by different events (=*intersections*), (2) outcomes that occur in any of two or more events (=*unions*), (3) outcomes that are not part of a particular event (=*complements*), and (4) outcomes whose chance of occurrence depends on the occurrence of other outcomes (=*conditional probabilities*). The combination of these tools (and some diligent bookkeeping) allows us to associate every event with its probability, and thereby to specify the *probability distribution* for an experiment. Through the use of the probability distribution we will be able to explore a number of biological phenomena in which chance plays a crucial role.

2.6 *Problems*

1. Hand gestures play an important role in many cultures. For example, a raised index finger with all other fingers folded is a common gesture among sports fans in the U.S., signifying "We're no. 1!" Inappropriate use of this gesture may be annoying to those around one but is unlikely to cause serious problems. In contrast, in the United States a raised middle finger with all other

fingers folded is used to signify extreme displeasure, and use of this gesture in the wrong situation or wrong company can lead to unfortunate results. Let us assume that every culture possesses exactly one such dangerous hand gesture, selected randomly sometime in the distant past. Now the problem. You have just parachuted into the jungles of New Guinea, and when rescued by the natives you wish to greet them with a friendly wave of your hand. At random, you raise anywhere between zero and five fingers (keeping the rest folded) and extend your hand toward your rescuers. What is the probability that you have just commited a grave social faux pas and are thereby in danger of being shot?

2. You and your spouse intend to have four children. Your mother-in-law contends that because there is equal probability of having a boy or a girl in each birth, it is most probable that you will have an equal number of boys and girls among your four kids. Is she correct? Why or why not?

3. Gambler 1 wins if he scores at least one "1" (an ace) in six throws of a single die. Gambler 2 wins if he gets at least two aces in twelve throws of a single die. Which gambler is more likely to win? It may motivate you as you work through the math to know that this problem was first posed by Samuel Pepys and solved by Sir Isaac Newton in 1693 (Feller 1960).

4. Four deer are captured from a population of N deer, marked, and released back into the population. After a time lapse sufficient to ensure that the marked deer are randomly distributed among the population, five deer are captured at random from the population. What is the probability that *exactly one* of these recaptured deer is marked if

- $N = 8$
- $N = 15$
- $N = 25$
- $N = 10$
- $N = 20$
- $N = 30$

Graph your results of probability versus population size. Can you provide an intuitive explanation for the shape of this curve?

5. A box of one hundred screws contains ten screws that are defective. You pick ten screws at random from the box. What is the probability that all ten screws you have chosen are good? What is the probability that exactly one is defective?

6. How many people do you have to assemble in a room (people picked at random from the general population) before there is at least an even chance ($P = 0.5$) that at least two have the same birthdate (e.g., February 4 or September 9)? Assume that the year has 365 days; that is, don't worry about leap year. *Hint*: Check your answer by using the same method to calculate the probability that two people will have the same birthday if there are 366 people in the room.

7. You are a contestant on a popular game show that allows you to choose from among three doors, each of which hides a prize. Behind one of the doors is the vacation of your dreams. Behind each of the other two doors is a block of moldy cheese. You make your selection and tell the host. In response, the host (who knows what is behind each door) opens one of the two doors you did not choose, revealing some odorous cheese. She then gives you a chance to change your mind and select the unopened door you did not choose originally. Would it be an advantage to switch? In other words, what is your probability of winning the vacation if you keep your original choice? What is your probability of winning if you switch? (*Note*: This puzzle stumped a number of mathematicians when it was posed in a newspaper in 1991; see Tierney 1991; Hoffman 1998.)

8. Suppose the game show in question 7 had four doors instead of three. If all rules of the game are otherwise the same, should you switch doors?

9. You live in a town of $n+1$ people, and are interested in the dynamics of rumors. You start a rumor by telling it to one other person, who then picks a person at random from the town and passes the rumor on. This second person likewise picks a recipient at random, and so forth. What is the probability that the rumor is told k times before it comes full circle and is repeated to you? What is the probability that the rumor is told k times before it is repeated to anyone? Work the problem again, but assume that each time the rumor is passed on, it is told to a group of N randomly chosen individuals. (This problem was borrowed from Feller 1960.)

10. Who's the father? A mare is placed in a corral with two stallions. One of the stallions is a champion thoroughbred racehorse worth millions of dollars. The other stallion looks similar but did not have such a distinguished racing career. The mare becomes pregnant and produces a colt. If the colt was fathered by the thoroughbred, he is worth a lot of money.

a. From the information given so far, what is the probability that the colt was fathered by the thoroughbred?

b. Suppose the thoroughbred has a relatively rare genetic marker on his Y chromosome that only occurs in 2% of horses. You know nothing about the genetics of the second stallion. You test the colt and find he also carries the rare marker. What is the probability the colt is the son of the thoroughbred given that he has the genetic marker?

c. Suppose the mare recently spent time roaming free on the range. During this time she was exposed to 998 other stallions who also could be the father of the colt. Now what is the probability that the colt is the son of the thoroughbred, given that he has the genetic marker?

d. What are the implications of this exercise to human legal trials where genetic markers are used to identify potential suspects?

3

Discrete Patterns of Disorder

In the last chapter we developed a number of rules to help us explore chance events. All of these rules were devised to allow us to enumerate the probability distribution for a random experiment. By being able to list all of the possible outcomes of an experiment and to associate each outcome with a probability, we have an invaluable tool; as we will see, knowledge about probability distributions allows us to solve many types of problems. But there is one issue that challenges our arsenal of techniques—large sample spaces. We have focused so far on relatively simple experiments with a small number of outcomes. But there are many other types of problems where the number of possible outcomes from an experiment is enormous. In such cases, estimating the probability distribution through repeated random experiments would be a daunting or even impossible task. Fortunately, we can gain insight into such complex sample spaces using theoretical distributions. Practical experience and mathematical insight have shown that a few theoretical probability distributions pop up repeatedly in a wide variety of chance events. By understanding the logic behind these common distributions, we greatly simplify the task of studying phenomena whose outcomes are uncertain. In this chapter, we develop approaches for dealing with discrete probability distributions with large, or even infinite, sample spaces. In chapter 4 we will expand our repertoire to include continuous sample spaces.

3.1 *Random Variables*

We begin by returning to the topic of wrestling matches between fish. When a fringehead loses a wrestling match with an intruder, it becomes homeless. After directing a few sarcastic thoughts at the home-wrecker, the displaced fish must move on to locate a new shelter. Searching the kelp forest for a new home costs the fringehead time that it could otherwise spend gathering food or attracting mates. Moreover, while the sarcastic fringehead is wasting time searching for a new shelter, it may also be more vulnerable to its own predators. Losing a wrestling match can be costly indeed.

Suppose that a displaced sarcastic fringehead must spend approximately three hours searching for a new shelter. Let's also assume that each day involves four encounters with invading fringeheads. Two questions then come to mind:

1. How much time will a fringehead waste in a typical day searching for a new shelter? This "average" value, summed over time, will determine how much wasted effort the fish expends in its life and is an index of its increased risk of being eaten.

2. How variable will the wasted time be from day to day? This question could be important, for instance, in the evolution of gut design. If there is little variability from one day to the next in the time available for foraging (and therefore in the amount of food consumed), natural selection might act to limit intestines to a size sufficient only to accommodate that specific daily consumption. If, instead, foraging time varies drastically from day to day, natural selection may favor a larger gut that can accommodate the occasional feast.

This general type of problem is very common. You have some unpredictable event (a wrestling match), and each outcome has a number associated with it (success, s : 0 hours of wasted time, failure, f : 3 hours wasted). There are a large number of stochastic processes that can be described in a similar form. For example, the chance event of a predator capturing a specific prey item yields a certain number of net calories, the random event of a bee pollinating a specific flower produces a certain number of fertilized embryos, or the random event of a wave crashing on the shore removes a certain number of mussels from the rocks.[1] These are all particular instances of something that in probability theory is called a *random variable*. A random variable is essentially a mathematical

---

[1] We were tempted to include in this list the chance event of a dropped piece of toast falling butter side down, an occurrence that results in the frustrating cost of mopping the floor. But is falling toast a random event? It is common lore that toast invariably lands butter side down—an oft-cited example of Murphy's Law. British physicist Robert Matthews used to think this was nonsense. He believed that by chance there would be a roughly equal opportunity for the toast to land butter side up or down. He suspected that the reason people think toast lands butter side down more often is that we tend to focus on these negative outcomes more than the innocuous butter-up landings. Like any good scientist, however, Matthews decided to test his intuition with experiments. Much to his surprise, he found that well more than half the time toast launched from a dining room table did indeed land butter side down. So much for intuition. To account for these surprising results, he explored the physics of falling toast (Matthews 1995, 1997). It turns out that when a piece of toast falls off a table, a gravitational torque causes the toast to rotate at a particular angular velocity. Given the height of most tables, there is only enough time for the toast to twirl approximately 180°; hence the probable result of landing butter side down. This finding offers a simple solution to the problem of cleaning up after fallen toast: eat on a much taller table! If toast slides off a 10-foot-high table, it has enough time to rotate 360°, thereby landing butter side up nearly every time.

TABLE 3.1 Wasted Time as a Function of the Outcome of Four Wrestling Matches

Outcome	Wasted Time	Outcome	Wasted Time
(s, s, s, s)	0 hr	(s, f, f, s)	6 hr
(s, s, s, f)	3 hr	(f, s, f, s)	6 hr
(s, s, f, s)	3 hr	(f, f, s, s)	6 hr
(s, f, s, s)	3 hr	(s, f, f, f)	9 hr
(f, s, s, s)	3 hr	(f, s, f, f)	9 hr
(s, s, f, f)	6 hr	(f, f, s, f)	9 hr
(s, f, s, f)	6 hr	(f, f, f, s)	9 hr
(f, s, s, f)	6 hr	(f, f, f, f)	12 hr

function that maps each outcome in the sample space to a particular value. We can then treat these values in the same way we have previously treated outcomes.

For example, in the case at hand, our random variable is the time wasted searching for shelters. If we assume exactly four wrestling matches a day, the sixteen possible outcomes (and the associated searching times) are as shown in table 3.1.

The amount of time wasted can be anywhere from 0 to 12 hours. Given this range of values, how much time is a fringehead likely to waste in a day? Being able to enumerate the sample space for the random variable is not enough— the answer to this question also depends on the *probability* of the different outcomes. If all the outcomes have an equal probability of occurring, we expect the average time wasted to be somewhere near the middle of the range simply because there are many more outcomes that waste 3, 6, or 9 hours than there are outcomes that waste 0 or 12 hours. But the different outcomes in the sample space could easily have very different probabilities. For example, the fringehead in question could be much larger than his neighbors, which would make its chance of losing any wrestling match relatively small. As a result, outcomes with many successes would be much more likely than outcomes with many failures, and our estimate of the time wasted in an "average" day would decrease. In other words, before we can predict what will happen in a typical day, we need to (1) enumerate the sample space of possible values, and (2) assign a probability to each value. In other words, we need to know the *probability distribution* for the experiment. How to do this is the basic goal of the next two chapters.

3.2 *Expectations Defined*

We set off toward that goal by developing a mathematical description of what we mean by "average." Given a list of k numbers (each the value of a random variable: $x_1, x_2, x_3, \ldots x_k$), every school child understands how to calculate the arithmetic average, $\overline{X}$. You sum up all the values and divide this sum by the number of numbers:

$$\overline{X} = \frac{\sum_{i=1}^{k} x_i}{k}. \tag{3.1}$$

For example, your test average in a course is the sum of your individual test scores (the values of the random variable) divided by the number of tests. An alternative but equivalent way to calculate the average of a series of k numbers is to multiply each individual value by $1/k$ and then sum the products:

$$\overline{X} = \sum_{i=1}^{k} \left(x_i \frac{1}{k} \right). \tag{3.2}$$

This form requires more work to calculate, but it provides a better conceptual feel for what the average is. If there are k values in the series, then $1/k$ is the frequency of occurrence of each individual number within the list. The sum of the quantity [the product of each value and its frequency of occurrence] gives the average.

This formula for calculating the average has the second advantage of making it easy to deal with situations where some of the values in the list are repeated. For example, in the series $(1, 2, 2, 2, 3, 4, 5, 5, 5, 5)$ there are ten entries (so $k = 10$) but only five different numbers; "2" appears three times and "5" four times. From our definition (eq. 3.2), we can calculate the average of this list:

$$\overline{X} = \left(1 \times \frac{1}{10} \right) + \left(2 \times \frac{3}{10} \right) + \left(3 \times \frac{1}{10} \right) + \left(4 \times \frac{1}{10} \right) + \left(5 \times \frac{4}{10} \right)$$
$$= 3.4. \tag{3.3}$$

Stated in a more general form,

$$\overline{X} = \sum_{\text{all } x} x \left(\frac{n_x}{k} \right), \tag{3.4}$$

where n_x is the number of entries that have value x. Values that occur more frequently (that is, values for which n_x is relatively large) have a greater impact on the overall average.

The logic we have used here to calculate $\overline{X}$ (the average of a specific list of values) can be extended to calculate another, related parameter, μ_X, the value we would expect to see from some chance event. Recall from chapter 2 that in the limit of a very large number of trials, the *probability* of an outcome is defined as the frequency of occurrence of that outcome. To put it another way, if we have a large number of trials (k, for instance) and a particular outcome happens a certain number of times (n_x, for example), the probability of that outcome is $P(x) = n_x/k$. Thus, *provided that k is very large,*

$$\mu_X = \sum_{\text{all } x} x P(x). \tag{3.5}$$

The function defined here, a weighted average of outcome values with weights defined by the probabilities that outcomes occur, is called the *expectation* of the random variable. It is directly analogous to $\overline{X}$, the average of a group of numbers (eq. 3.2), except that the probability of getting a particular value replaces the frequency of occurrence.

This particular formulation is very important. There will be many occasions when we want to calculate the expectation of something, and it behooves us to carefully define the expectation function, $E(X)$:

DEFINITION 2 *The* expectation *of a discrete random variable X, written $E(X)$, is defined by*

$$E(X) \equiv \sum_{\text{all } x} x P(x). \tag{3.6}$$

Note that this is a general definition. So far we have used it to calculate μ_X, in which case each x in the formula is a discrete datum. It is possible to take the expectation of other values, as we will demonstrate shortly when we discuss the variance.

The expectation of a variable is what most people would intuitively think of as the average (or *mean*). For instance, in our example here, the expectation of wasted time is the value that a fringehead should expect to waste in a typical day. There is, however, an aspect of this definition that may not be intuitive. Because the expectation is calculated using probabilities, it is an idealized quantity. In other words, it is our best guess as to how a variable will behave "on average." In the real world (where, for instance, wrestling matches do not come in infinite numbers), the expectation will commonly differ from what really happens in any particular instance. Only for a very large number of experiments are expectation and reality likely to converge. This difference between the expectation (an ideal) and the average that we might actually encounter (an empirical measure) should be kept in mind as this chapter progresses. In order to keep these two concepts straight we have given them separate symbols. The expectation of a variable is denoted by $\mu(X)$ or (in streamlined form) μ_X, the empirical average by $\overline{X}$, where in each case X is the random variable.

There are four corollaries to Definition 2 that we should note. The first was tacitly implied above: if our random variable X is a constant, by definition it does not vary, and every experiment involving this variable will have exactly the same outcome. In this case, the expected value of the variable is simply its constant value.

COROLLARY 2.1 *The expectation of a constant is that same constant.*

Second, consider two random variables, A and B, each a function of X. For example, A might be $2X$ and B might be $\sqrt{X}$. With just a little algebra, we can derive an expectation for the sum of these random variables. By definition,

$$E(A + B) = (a_1 + b_1) \cdot p_1 + (a_2 + b_2) \cdot p_2 \ldots , \qquad (3.7)$$

where (in this example) $a_i = 2x_i$, $b_i = \sqrt{x_i}$, and p_i is the probability associated with $(a_i + b_i)$. By distributing the probabilities, we see that

$$E(A + B) = (a_1 \cdot p_1) + (b_1 \cdot p_1) + (a_2 \cdot p_2) + (b_2 \cdot p_2) \ldots$$
$$= (a_1 \cdot p_1 + a_2 \cdot p_2 \ldots) + (b_1 \cdot p_1 + b_2 \cdot p_2 \ldots). \qquad (3.8)$$

Again applying the definition of the expectation, we conclude that

$$E(A + B) = E(A) + E(B). \qquad (3.9)$$

In other words:

COROLLARY 2.2 *The expectation of the sum of random variables is equal to the sum of the expectations of the individual variables.*

Now consider the random variable A and a constant C. By definition,

$$E(C \cdot A) = C \cdot a_1 \cdot p_1 + C \cdot a_2 \cdot p_2 \ldots$$
$$= C \cdot (a_1 \cdot p_1 + a_2 \cdot p_2 \ldots)$$
$$= C \cdot E(A). \qquad (3.10)$$

Thus:

COROLLARY 2.3 *The expectation of the quantity [the product of a constant and a variable] is the product of the constant and the expectation of the variable.* $\qquad (3.11)$

And, finally, consider the expectation of the product of two random variables, X and Y. If X and Y are not independent, this expectation is messy. However, *if X and Y are independent*, the result is pleasingly simple. The calculation is as follows. By definition,

$$E[X \cdot Y] = \sum_{\text{all } X} \left[\sum_{\text{all } Y} (X \cdot Y) \cdot P(X \cap Y) \right], \qquad (3.12)$$

where $P(X \cap Y)$ is the probability that both X and Y occur. But if X and Y are independent, $P(X \cap Y) = P(X) \cdot P(Y)$ (rule 6 from chapter 2). So,

$$E[X \cdot Y] = \sum_{\text{all } X} \left[\sum_{\text{all } Y} [X \cdot P(X)] \cdot [Y \cdot P(Y)] \right]. \tag{3.13}$$

For any given X, $X \cdot P(X)$ is constant for all Y, so

$$E[X \cdot Y] = \sum_{\text{all } X} \left\{ [X \cdot P(X)] \cdot \sum_{\text{all } Y} [Y \cdot P(Y)] \right\}. \tag{3.14}$$

But $\sum_{\text{all } Y} [Y \cdot P(Y)]$ is the expectation of Y. Thus,

$$\begin{aligned} E[X \cdot Y] &= \sum_{\text{all } X} [(X \cdot P(X) \cdot E(Y)] \\ &= E(Y) \cdot \sum_{\text{all } X} [X \cdot P(X)] \\ &= E(X) \cdot E(Y). \end{aligned} \tag{3.15}$$

In other words:

COROLLARY 2.4 *If two random variables are independent, the expectation of their product is equal to the product of their expectations.*

3.3 *The Variance*

We now explore our second question regarding fringehead displacement, that of the variation in time wasted from day to day. As we have noted, because there is a wide range of possible values, extreme days might affect fringehead evolution. As we warned you above, in any given day, the actual time wasted (T_{waste}) can differ from the expected time wasted $E(T_{\text{waste}})$. Our second question relates, then, to the expected value of this *deviation* from the long-term mean. In other words, what is $E[T_{\text{waste}} - \mu(T_{\text{waste}})]$? This is easy to solve. From our second corollary,

$$E[T_{\text{waste}} - \mu(T_{\text{waste}})] = E(T_{\text{waste}}) - E[\mu(T_{\text{waste}})]. \tag{3.16}$$

Now, $E(T_{\text{waste}}) = \mu(T_{\text{waste}})$. Furthermore, the average, $\mu(T_{\text{waste}})$, is a constant, and the expectation of a constant is that constant. So,

$$\begin{aligned} E[T_{\text{waste}} - \mu(T_{\text{waste}})] &= \mu(T_{\text{waste}}) - \mu(T_{\text{waste}}) \\ &= 0. \end{aligned} \tag{3.17}$$

In other words, the average deviation from the expectation is exactly 0. Does this mean that there will be no variation in time wasted? Of course not; we are, after all, dealing with a random process. On some days, the wasted time will be larger than $\mu(T_{\text{waste}})$, and on other days the time wasted will be smaller than $\mu(T_{\text{waste}})$. It's just that through time the positive differences are expected exactly to balance the negative differences. To get a measure of variability, we need to modify our formula so that the positive and negative differences from the mean do not cancel each other out.

There are a number of ways to accomplish this task. For example, we could take the absolute value of $T_{\text{waste}} - \mu(T_{\text{waste}})$ before calculating the expectation. An alternative approach (which turns out to have some advantageous mathematical properties) is to square the average deviation from the mean. Squaring the deviations makes them all positive. The expectation of the squared deviation of a random variable X from its mean is called the *variance*, written $\sigma^2(X)$, or (in streamlined form) σ_X^2:

DEFINITION 3 *The* variance *of a discrete random variable X is*

$$\sigma^2(X) \equiv E(X - \mu_X)^2. \tag{3.18}$$

Before we continue with our discussion of the variance, it will be useful to note a corollary of the definition given above. Using the formula for the expectation, we can rewrite the definition of the variance:

$$E(X - \mu_X)^2 = \sum_{\text{all } x} \left[(x - \mu_X)^2 P(x) \right], \tag{3.19}$$

where $P(x)$ is again the probability that a particular value of the variable x will occur. Expanding the squared term, we see that

$$\sigma_X^2 = \sum_{\text{all } x} \left[(x^2 - 2x\mu_X + \mu_X^2) \cdot P(x) \right]$$

$$= \sum_{\text{all } x} x^2 P(x) - 2\mu_X \sum_{\text{all } x} x P(x) + \mu_X^2 \sum_{\text{all } x} P(x). \tag{3.20}$$

Several parts of this expanded equation should look familiar. In particular, $\sum_{\text{all } x} x P(x)$ is the equation for the mean, μ_X, and we know that $\sum_{\text{all } x} P(x) = 1$ (rule 2, chapter 2). In addition, given our definition of the expectation, the term $\sum_{\text{all } x} x^2 P(x)$ is the average of the squared values of x, commonly referred to as the *mean square*, $\overline{X^2}$. Thus,

$$\sigma_X^2 = \overline{X^2} - 2\mu_X^2 + \mu_X^2 \tag{3.21}$$

$$= \overline{X^2} - \mu_X^2. \tag{3.22}$$

In other words,

> COROLLARY 3.1 *The variance is equal to the mean square minus the mean squared.*

In a standard statistics course, you may have encountered this corollary as one form of the "machine formula" for calculating the variance.

Note that since the variance is a squared deviation, it has units that are squared. If our values are measured in terms of hours or degrees or calories, the variance will have nonintuitive units that make it difficult to compare with the expectation. As a result, it is often helpful to describe the variation of a process not in terms of its variance, but rather in terms of the square root of the variance, a value known as the *standard deviation*, symbolized by $\sigma(X)$ or σ_X.

> DEFINITION 4 *The* standard deviation *of a random variable X is defined as the square root of the variance of X.*

Returning to our practical problem of calculating the expected daily amount of time wasted by the fringehead and how variable the time is likely to be, we now know how to calculate estimates for both. For the expectation (the idealized average) we take the sum over all the outcomes in our sample space of the product of the value associated with each outcome (i.e., the time wasted) and the probability of that outcome's occurrence. Having calculated the expectation, we can then calculate the variance or standard deviation. We only have one problem: We do not yet know the probability associated with the different outcomes. This brings us back to the topic of probability distributions. How can we deduce the probability distribution of $\mathcal{T}_{\text{waste}}$? To do so, we first explore the concept of Bernoulli trials.

3.4 *The Trials of Bernoulli*

The simplest nontrivial experiment is one that has only two possible outcomes. Examples would be getting a heads or a tails on the flip of a coin, having a male or a female baby, having a red blood cell go left or right at an arterial junction, or winning or losing a battle for a mate. In probability theory, these types of experiments are called *Bernoulli trials*, in honor of Jacob Bernoulli.[2] To examine a random variable associated with a Bernoulli trial, we have to associate a value with each outcome. For simplicity, one of the possible outcomes typically is considered a success and is assigned a value of 1. The other outcome is considered a failure and is assigned a value of 0. (We will return to the

---

[2] Jacob Bernoulli's real-world trial was his brother, Johannes. Both Bernoullis were mathematicians and very competitive. See Bell (1937).

more general case where the values associated with the two outcomes can be any number.) The choice of which outcome is called a success in a Bernoulli trial is often arbitrary; it may have nothing to do with winning or losing.

The probability distribution for a Bernoulli random variable is very simple. If the probability of a success is p, then the probability of a failure is $q = 1 - p$, and these two values comprise the entire distribution.

Knowing the probability distribution for a Bernoulli trial, can we calculate its expectation and standard deviation? Consider a Bernoulli random variable, B. Using the definition of an expectation (see Definition 2), we have

$$E(B) = \mu_B = p \cdot 1 + q \cdot 0 = p. \tag{3.23}$$

The expected value of a Bernoulli random variable is nothing more than the probability of a success.

Let's move on to the variance and standard deviation of a Bernoulli random variable. In a fraction p of the trials we observe a success, giving a value of 1 in each case. In the remaining fraction q of the trials we observe a failure, giving a value of 0. Therefore, combining the definition for a variance (see Definition 3) and the definition of the expectation (see Definition 2), we have

$$\sigma^2(B) = p \cdot (1 - \mu_B)^2 + q \cdot (0 - \mu_B)^2. \tag{3.24}$$

Now,

$$\mu_B = p \text{ (see eq. 3.23)}, \tag{3.25}$$

so,

$$\sigma^2(B) = p(1 - p)^2 + q(0 - p)^2. \tag{3.26}$$

But

$$q = 1 - p, \tag{3.27}$$

so,

$$\sigma_B^2 = pq^2 + qp^2. \tag{3.28}$$

Factoring, we get

$$\sigma_B^2 = pq(p + q). \tag{3.29}$$

Again recalling that $p + q = 1$, our final result is

$$\sigma_B^2 = pq. \tag{3.30}$$

The variance of a Bernoulli random variable is the product of the probability of a success and the probability of a failure. It has a maximum value of 0.25 when $p = q = 0.5$.

From the variance we can solve for the standard deviation,

$$\sigma_B = \sqrt{pq}. \tag{3.31}$$

3.5 *Beyond 0's and 1's*

Most introductory books on probability theory stop at this point and assume you can easily make the jump to situations where the two outcomes have values other than 0 and 1. Our experience suggests otherwise. For example, in our fringehead experiment, the value of a wrestling win is 0 hours wasted, whereas the value of a wrestling loss is 3 hours wasted. Can you look at the formulas for the expectation and standard deviation of a Bernoulli random variable and readily predict how they change with outcome values other than 0 and 1? We didn't think so.

Here is a more general derivation of the expectation and standard deviation for Bernoulli trials with the two outcomes having values V and W, respectively. First, the expectation. From Definition 2,

$$E(B) = \mu_B = (pV) + (qW). \tag{3.32}$$

This is not too bad. We can use this result to get the first piece of the answer to our question of how much time a fringehead will waste searching for new shelters any given day. If we let p be the probability of a fringehead winning a wrestling match with an intruder, we can calculate the average amount of time wasted from a single wrestling match (T_{match}). In this case, $V = 0$ hours and $W = 3$ hours. Hence,

$$\mu(T_{\text{match}}) = p \cdot 0 + q \cdot 3$$
$$= 3q \text{ hours.} \tag{3.33}$$

So, if a large fringehead has a 0.25 probability of losing a random wrestling match, it would, on average, expect to waste 0.75 hours per match.

What about the variation in time wasted from match to match? This derivation will get a bit messy for a while, but it will work out nicely in the end. Let's start with the general formula for the variance of a Bernoulli random variable. Again, we use two outcomes with values V and W:

$$\sigma_B^2 = p \cdot (V - \mu_B)^2 + q \cdot (W - \mu_B)^2. \tag{3.34}$$

Substituting in our formula for μ_B (eq. 3.32), we have

$$\sigma_B^2 = p \cdot [V - (pV + qW)]^2 + q \cdot [W - (pV + qW)]^2. \qquad (3.35)$$

With a little rearrangement, we can simplify this a lot:

$$\begin{aligned}
\sigma_B^2 &= p \cdot [(1 - p)V - qW]^2 + q \cdot [(1 - q)W - pV]^2 \\
&= p \cdot [qV - qW]^2 + q \cdot [pW - pV]^2 \\
&= p \cdot [q(V - W)]^2 + q \cdot [p(W - V)]^2 \\
&= pq^2(V - W)^2 + qp^2(W - V)^2 \\
&= (pq^2 + qp^2) \cdot (V - W)^2 \text{ since } (V - W)^2 = (W - V)^2 \\
&= pq(q + p) \cdot (V - W)^2. \qquad (3.36)
\end{aligned}$$

Recalling that $p + q = 1$, we see that

$$\sigma_B^2 = pq \cdot (V - W)^2. \qquad (3.37)$$

From here we can easily get the standard deviation:

$$\sigma_B = \sqrt{pq} \cdot |V - W|. \qquad (3.38)$$

Thus, the standard deviation of the generalized form of the Bernoulli random variable is just the standard deviation from the special case, $\sqrt{pq}$, times the absolute value of the difference between the values of the two outcomes. The more different the values of the two outcomes are, the more variable the observed results will likely be—an intuitively pleasing result.

We can use this formula to estimate the standard deviation for a single fringe-head wrestling match. Again, if we assume a 0.25 probability of losing (that is, $q = 0.25$), the standard deviation of the time wasted is

$$\begin{aligned}
\sigma(T_{\text{match}}) &= \sqrt{\frac{3}{4} \cdot \frac{1}{4}} \cdot |0 - 3| \\
&= \sqrt{\frac{3}{16}} \cdot 3 \cong 1.3 \text{ hours.} \qquad (3.39)
\end{aligned}$$

3.6 $\sum$ *Bernoulli = Binomial*

The concept of a Bernoulli random variable is quite simple, and (as we have seen) this random variable is a good analogy for a single contest with two outcomes. Knowing the probability distribution for this experiment allows us to calculate the expected time that is wasted from a single match. But how can we calculate the amount of time wasted by a fringehead through the course of the four matches that occur in a typical day? If one wrestling match is a Bernoulli random variable, we need a new variable that is the sum of multiple Bernoulli variables, one from each wrestling match that we expect in a day.

Let's consider this new random variable. Each wrestling match is a Bernoulli trial with *two* possible outcomes: success (defined as a win) and failure (defined as a loss). The total number of successes in a single Bernoulli trial is either 0 or 1. In two matches there can be *three* outcomes: two losses, two wins, or a win and a loss. Each outcome can again be defined in terms of the total number of successes, in this case 0, 1, or 2. The logical extension of this process is that given n total matches, there are $n + 1$ possible outcomes, and these outcomes can be associated with the integers from 0 to n.

Having thus defined our variable, we now ask an important question: What does the probability distribution of this new variable look like? To define the probability distribution, we need to know the probability of getting each number of successes in n trials. That is, if we have n Bernoulli trials (e.g., wrestling matches), what is the probability of getting exactly i successes?

Let's start with the extreme case of zero successes—every wrestling match ends with the fringehead looking for a new home. If the probability of failure on any given trial is q, the probability of an outcome with zero successes in n wrestling matches is q^n. This result is just a consequence of rule 6 from chapter 2—the Product Rule: if two or more events are independent, the probability that all events occur is the product of their individual probabilities.

We can use similar logic to calculate the probability of outcomes that are mixtures of successes and failures. For example, what is the probability of getting a win in the first match followed by $n - 1$ losses? Since the events are independent, the win occurs with probability p and the $n - 1$ losses occur with probability q, so the event occurs with probability pq^{n-1}. In fact, it is simple to calculate the probability of any specific chain of wins and losses by the same approach. If the chain includes i successes and $n - i$ failures, the probability of that specific chain of events is $p^i q^{n-i}$.

Now, $p^i q^{n-i}$ is the probability for a particular succession of wins and losses with a total of i successes. However, there could be multiple outcomes with i successes. For example, if we return to our table of daily wrestling matches (table 3.1), there are four outcomes with a single "win" ($sfff$, $fsff$, $ffsf$, $fffs$). Each of these four outcomes occurs with probability pq^3. Therefore, the overall probability of getting exactly one success in four trials is the probability of an individual outcome with one success ($= pq^3$) times the number of possible outcomes with one success ($=4$). To put this into mathematical terms, let S be a random variable equal to the number of successes (i) in n trials. Thus, the overall probability of getting i successes is

$$P(S = i) = p^i q^{n-i} \cdot [\text{number of outcomes with } i \text{ successes}]. \qquad (3.40)$$

The function $P(S = i)$ defines the probability distribution of the random variable S, and can thus be used to describe the probability distribution for our daily wrestling matches. But this probability distribution is not nearly as simple as the one for a single Bernoulli random variable. The term [number of outcomes with i successes], complicates things considerably.

One way we can calculate the number of outcomes with i successes is to create a table of the sample space and arduously total up the appropriate number of outcomes. This is what we did for the random variable that describes the fringehead wrestling matches. For cases with a small number of trials, it is possible to use this counting technique to estimate the number of outcomes with i successes, but this approach becomes quite tedious as the number of trials increases. Imagine trying to create a table of outcomes for fringehead wrestling matches in a month. Since there are two possible outcomes for each Bernoulli trial, n trials will produces 2^n possible outcomes. With 120 matches there would be $1.33 \cdot 10^{36}$ cells in your table! It would be very useful to have a general solution for finding the number of outcomes with i successes.

3.6.1 PERMUTATIONS AND COMBINATIONS

Now, as we have seen, there are multiple outcomes with i successes because the successes come in different *orders* during the n Bernoulli trials. What we need, then, is a formula for how many ways we can order i successes in n trials (in the jargon of probability theory, the number of *combinations*). This is often written as $\binom{n}{i}$. To calculate $\binom{n}{i}$, we need to "place" our i successes into our n trials, a task analogous to placing balls in boxes, one ball per box.

If we have i balls and n boxes, there are n options for placing the first ball in a box. Once one box is occupied, there are only $n - 1$ options for placing the second ball in an empty box. Continuing with this logic, it is clear that the total number of options available to you in placing the i balls is

$$n \cdot (n - 1) \cdot (n - 2) \cdot \ldots \cdot (n - i + 1). \qquad (3.41)$$

For example, if you have three balls and five boxes in which to arrange them, $(n - i + 1) = (5 - 2) = 3$, and you have

$$5 \cdot 4 \cdot 3 = 60 \qquad (3.42)$$

arrangements.

Now, this formula bears a resemblance to the formula for a factorial:

$$n! \equiv n \cdot (n-1) \cdot (n-2) \cdot \ldots \cdot 1. \tag{3.43}$$

(For future reference, note that $0! \equiv 1$.) The number of options (eq. 3.41) includes all of the terms in $n!$ before $(n-i) \cdot (n-i-1) \cdot \ldots \cdot 1$. If you look carefully at these missing terms, you will see that they comprise the formula for $(n-i)!$. For example, given $n = 5$ boxes and $i = 3$ balls, $(n-i) = 2$, and $2!$ is $2 \cdot 1$. Thus,

$$\frac{5!}{2!} = \frac{5 \cdot 4 \cdot 3 \cdot 2 \cdot 1}{2 \cdot 1} = 5 \cdot 4 \cdot 3. \tag{3.44}$$

Therefore, if we divide $n!$ by $(n-i)!$, we are left with the formula above (eq. 3.41) for the number of options we had in placing i balls into n boxes. That is,

$$\text{number of different options} = \frac{n!}{(n-i)!}. \tag{3.45}$$

Let's check to see if this formula gives us the estimate of $\binom{n}{i}$ that we seek. Recall from our fringehead experiment (table 3.1), that there were six ways to obtain two wins in four wrestling matches. If we use the formula above, we get

$$\frac{4!}{(4-2)!} = \frac{4 \cdot 3 \cdot 2 \cdot 1}{2 \cdot 1} = 12. \tag{3.46}$$

This is *not* the correct answer. The flaw in our logic is that many of the arrangements we counted while placing balls in boxes generate exactly the same ultimate distribution of balls in boxes, it's just that the *order* in which they were placed in the boxes differs.

For example, suppose the first ball went into box 1 and the second ball went into box 5. This gives exactly the same end result as the first ball going into box 5 and the second ball going into box 1. The formula for the number of different options, $n!/(n-i)!$, unfortunately distinguishes the order that balls were placed into the boxes, and counts them separately. But for our question regarding the variability in wasted time, a success is a success, and the order is irrelevant. As a result, we have to correct our estimate for the number of outcomes with i successes by (in effect) reducing it by the number of ways we can put the i successes into i specific boxes.

Fortunately, this is a simple task. Suppose that after putting our i balls into the n boxes, we take the boxes with balls and remove the balls. How many different ways can we put the i balls back into these i boxes? Well, there are i ways to put the first ball into a box, $(i-1)$ ways to put the second ball into an unoccupied box, etc. In other words, by the same logic we used before, we can conclude that there are $i!$ different ways (termed *permutations*) to put the balls back into the same boxes. This then is the number of identical patterns we generated by our overly simplistic technique above.

As a result, if we divide our calculated total number of options above by $i!$, the number of permutations, we get the correct estimate for the number of combinations $\binom{n}{i}$:

$$\binom{n}{i} = \frac{n!}{i!(n-i)!}. \tag{3.47}$$

For example, when placing two balls in four boxes, there are

$$\frac{4!}{2!2!} = \frac{4 \cdot 3 \cdot 2 \cdot 1}{(2 \cdot 1) \cdot (2 \cdot 1)} = 6 \tag{3.48}$$

combinations, the correct answer. With this formula in hand, we immediately arrive at a very important result. *The probability distribution for a random variable defined by the sum of n independent Bernoulli trials is*

$$P(S = i) = \frac{n!}{i!(n-i)!} \cdot p^i q^{n-i}. \tag{3.49}$$

In other words, given the overall number of trials (n) and the probability of success on any one trial (p), this formula allows us to associate each number of successes (i) with its corresponding probability. When you think about it, this is a really nifty result. No longer do we have to go through the tedious procedure of enumerating the sample space and noodling out what probability to assign to each outcome. We now have a compact formula that does most of the work for us. This distribution is called the *binomial distribution*, and a random variable with this probability distribution is called a *binomial random variable*.[3] Some sample plots of binomial distributions are plotted in figure 3.1.

Let's use the formula for the binomial distribution to calculate some probabilities that will be of use in chapter 6. Consider two spawning salmon, the last remaining members of the salmon population of a creek. Each adult produces a multitude of haploid gametes, each carrying one allele for a particular gene (the gene for specific form of enzyme, for instance). Gametes are mixed at random and many eggs are fertilized to form diploid larvae, but only two survive. What can we predict about the allelic composition of these two surviving larvae?

Let's take a simple example and assume that half the sperm and half the eggs have type A alleles, the remaining gametes have a alleles. Each of our larvae

---

[3] The term "binomial" stems from the fact that the probability values in this distribution can alternatively be calculated from the expression $(p+q)^n$. Because the term in parentheses contains two variables, it is known as a *binomial*. If n is large, it is much easier to calculate the probability distribution using eq. (3.49) than using this expansion directly.

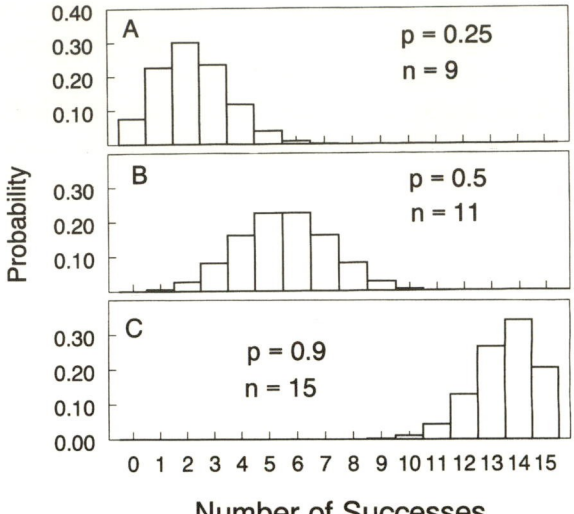

Number of Successes

FIG. 3.1 Examples of the binomial distribution. The shape of the distribution varies depending on the number of trials (n) and the probability of success (p).

has two alleles at this locus, for a total of four. What is the probability that the allelic composition of the larvae is the same as that of the parents? In other words, what is the probability that the alleles in the surviving larvae consist of exactly two A and two a alleles?

As we have suggested, this question can be answered using the binomial distribution. To set up the problem, we define the random "choice" of an A allele as a success. Because A alleles form half the allelic population, the probability of randomly choosing an A allele is $p = 0.5$. The probability of obtaining half A alleles and half a alleles in the four alleles in our two larvae is thus the probability of getting two successes in four trials where the probability of success is 0.5. In mathematical terms,

$$P(\text{half } A \text{ alleles}) = \binom{4}{2} p^2 q^{4-2} = 0.375. \tag{3.50}$$

There is a 37.5% chance that the allelic composition of the larvae will be half A and half a, the same as the parents. Conversely, there must be a 62.5% chance that the allelic composition of the young is *different* from the parents. It is this chance that the gene pool of the young differs from that of the parents that drives genetic drift, a phenomenon we return to in chapter 6.

In addition to the probability distribution for a binomial random variable, we can calculate other useful quantities. For instance, let's calculate the expectation, the value we need to solve our problem regarding the average time wasted due to fringehead wrestling matches. We already have an estimate for the expected amount of time a fish would waste as a result of a single match, but we need to know the amount of time wasted on average in a day with multiple fights.

For a binomial random variable (S), the expected value is calculated from the expectation of the underlying Bernoulli random variables. Because the Bernoulli trials that make up the binomial variable are independent, the expectation of the binomial is simply the sum of the expectations of the n Bernoulli trials. Using the general form of the Bernoulli expectation with values V and W, and multiplying by n, we see that the expected value for the binomial is

$$\mu(S) = n(pV + qW). \tag{3.51}$$

This means that the fringehead who has four fights a day with a 0.25 probability of losing each match can expect to waste

$$\mu(\mathcal{T}_{\text{waste}}) = 4 \cdot [(0.75 \cdot 0) + (0.25 \cdot 3)]$$
$$= 3 \text{ hours per day.}$$

This then is the value a fringehead can expect to waste each day searching for new shelters.

So far, so good, but how variable is the wasted time likely to be? The variance of the binomial turns out to be equally simple. Because the trials are independent, the variance of the sum (that is, the variance of the binomial random variable) is simply the sum of the variances of the Bernoulli trials. But it will take some work to show this. Let's begin by returning to our definition of a variance (see Definition 3) and applying it to the sum of two random variables, X and Y:

$$\sigma^2(X + Y) = E[(X + Y) - \mu(X + Y)]^2. \tag{3.52}$$

We can expand the right-hand side by using our finding that the expectation of a sum is the sum of the expectations:

$$\sigma^2(X + Y) = E[(X + Y) - \mu_X - \mu_Y]^2. \tag{3.53}$$

Rearranging gives

$$\sigma^2(X + Y) = E[(X - \mu_X) + (Y - \mu_Y)]^2. \tag{3.54}$$

By squaring, taking the expectation term by term, and factoring, we find that

$$\sigma^2(X + Y) = E[X - \mu_X]^2 + 2 \cdot E[(X - \mu_X) \cdot (Y - \mu_Y)]$$
$$+ E[Y - \mu_Y]^2. \tag{3.55}$$

If you look carefully at the right side of this equation, two of the terms should be familiar. The first term, $E(X - \mu_X)^2$, is simply the variance of X. Similarly, the third term is the variance of Y. Therefore, our variance of a sum is equal to

$$\sigma^2(X + Y) = \sigma^2(X) + \sigma^2(Y) + 2 \cdot E[(X - \mu_X) \cdot (Y - \mu_Y)]. \tag{3.56}$$

If it were not for the nagging term on the right, our hopes for a simple expression for the variance of a sum would be realized.

Let's examine this last term to see what it represents. Notice that this term has some characteristics of our definition of a variance. It is the product of two deviations of a random variable from the expected value of that variable. Indeed, if both bracketed terms dealt with the same random variable, this would simply be a variance. Instead, this variance-like term includes variation from two different variables (X and Y), and it is called the *covariance* of X and Y. It measures the relationship between variation in two random variables. Positive covariance implies that whenever X is above its average value, Y has a tendency to be above its average as well. Similarly, if two variables are positively covariant and X is below its average, Y is likely to be below its average too. In contrast, a negative covariance implies that when X is above its average, Y is more commonly below its average, and when X is below its average, Y has a tendency to be above its average.

Returning to our problem, we ask what pattern of covariation we would expect between our random variables—a set of independent Bernoulli trials? The key word here is *independent*. Independence means that knowing a particular value of X tells us nothing about the value of Y we would expect to see (see chapter 2). This means that the expectation of the product in the covariance can be expanded to the product of two independent expectations (Corollary 2.4):

$$E[(X - \mu_X) \cdot (Y - \mu_Y)] = E[X - \mu_X] \cdot E[Y - \mu_Y]. \tag{3.57}$$

But, from eq. (3.17) we already know that the expected deviation of any random variable from its average is always zero. Therefore, the covariance of two independent random variables is always zero. As a result, the variance of the sum of two (or more) random variables is indeed the sum of the two (or more) variances, *as long as the variables are independent*:

$$\sigma^2(X + Y) = \sigma^2(X) + \sigma^2(Y). \tag{3.58}$$

Returning—finally—to our fringehead problem, we now know that because the variance of each Bernoulli trial is $pq(V - W)^2$ (see eq. 3.37), the variance of the sum of n Bernoulli trials (= binomial random variable, S) is

$$\sigma^2(S) = npq(V - W)^2. \tag{3.59}$$

For our example, in which $p = 0.75$, $V = 0$, and $W = 3$, the variance of the daily wasted time is

$$4 \times 0.75 \times 0.25 \times 3^2 = 6.75 \text{ hours}^2. \tag{3.60}$$

Yikes! Square hours! To talk about real time, we take the square root of this variance to arrive at the standard deviation. For a binomial probability distribution,

$$\sigma = \sqrt{npq} \cdot |V - W|$$

$$\cong 2.6 \text{ hours}. \tag{3.61}$$

In this particular example, the standard deviation of wasted time (2.6 hours) is only a bit less than the average wasted time (3 hours). Because the standard deviation is a substantial fraction of the mean, the fringehead should expect considerable variation from day to day.

If you have been carefully following our discussion to this point, the last sentence may sound a bit vague. Yes, the fish should expect "considerable variation" in the time it wastes, but what does this really mean? In other words, now that we have calculated the standard deviation, exactly what does it tell us?

Surprisingly little in this case. For example, we might ask the following question. How probable is it that in a given day the fish will waste time in excess of one standard deviation above the mean? That is, given a daily mean of 3 hours, and a standard deviation of 2.6 hours, how likely is it that the fish will waste more than 5.6 hours in a randomly chosen day?

Unfortunately, the answer cannot be had directly from a knowledge of the standard deviation alone. The probability associated with events at least one standard deviation above the mean depends on the shape of the particular probability distribution. To see why, consider the case we have just explored. Here p (the probability of winning) is 0.75, $V = 0$, $W = 3$, and the probability distribution is as shown in figure 3.2A. The distribution is asymmetrical, with most of the probability associated with the lower values of T_{waste}. To calculate the probability that $T_{\text{waste}} > \mu(T_{\text{waste}}) + \sigma(T_{\text{waste}})$, we sum the probabilities for all events that result in $T_{\text{waste}} > \mu(T_{\text{waste}}) + \sigma(T_{\text{waste}})$. This sum is $P = 0.26$; only 26% of days waste time in excess of the mean plus the standard deviation.

What happens if we assume instead that $p = 0.25$ rather than 0.75? The distribution changes shape (fig. 3.2B), and most of the probability is now associated with the higher values of T_{waste}. Despite the change in shape, we calculate exactly the same standard deviation for the distribution (see eq. 3.61). Working through the calculation, we find that for $p = 0.25$ $P[T_{\text{waste}} > \mu(T_{\text{waste}}) + \sigma(T_{\text{waste}})] = 0.32$, a slightly larger value than when $p = 0.75$. In

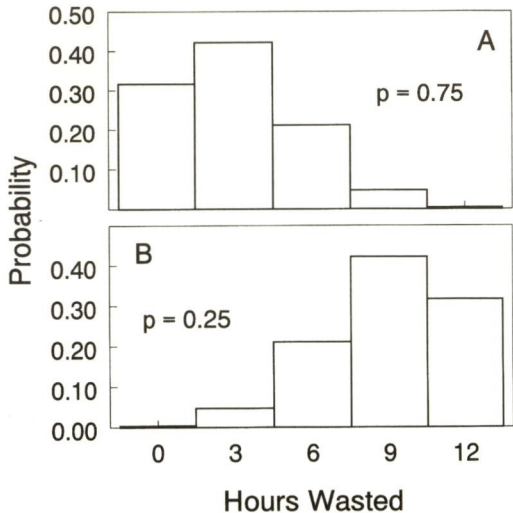

FIG. 3.2 The probability distribution of hours wasted in a day by a sarcastic fringehead. When the probability of success is high (A), most outcomes result in little time wasted. If the probability of success is low (B), most outcomes result in substantial wasted time. Note that even though these two distributions are skewed in opposite directions, their standard deviations are the same (eq. 3.61).

other words, *the probability associated with events at least one standard deviation away from the mean varies depending on the shape of the probability distribution.*

This leads us to an important conclusion. In general, the standard deviation is merely an index of the overall variability associated with a stochastic process. To relate the standard deviation to questions of probability, one has to connect the standard deviation to the particular *shape* of the probability distribution from which it was calculated.

Why, then, have we gone to all this effort to calculate σ? The answer will become apparent in chapter 5 when we explore the *normal distribution*. This family of probability curves *all have the same shape*. As a result, when dealing with normal distributions, knowledge about the standard deviation *can* be applied directly to questions about probability. It is primarily because normal distributions are found so often in stochastic processes that the standard deviation has practical utility.

3.7 *Waiting Forever*

Now that we have a random variable that describes a string of successes and failures (the binomial random variable), we can ask a special question: How long (on average) do we have to wait for the first success? This is akin to the question we asked in chapter 2: What is the probability that a fringehead will win its first wrestling match in the ith bout? In this case, however, we are not interested in a particular bout, but rather the average over all bouts.

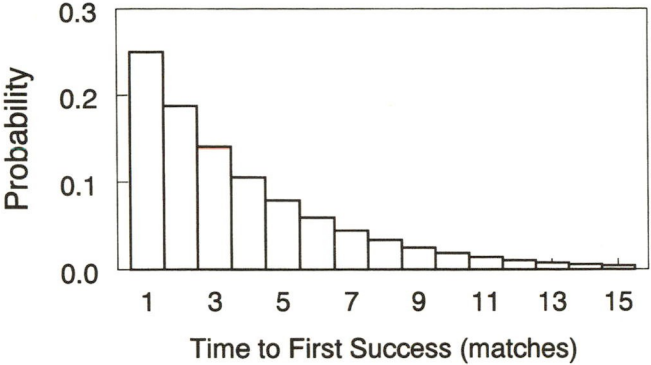

Fɪɢ. 3.3 An example of a geometric probability distribution: the probability that a fringehead will experience its first win in a particular wrestling match (eq. 3.62). Note that we have plotted only a small portion of the overall distribution, which extends to an infinite number of matches.

To answer this question, we need to find the expected value for a random variable (let's call it G) that describes a chance event characterized by a string of failures followed by a single success. This random variable is commonly called a *geometric random variable*.[4] Obtaining the first success on the nth trial requires $n - 1$ failures, each with probability $(1 - p)$, before the single success occurs with probability p. Since there is only one way to string together $(n - 1)$ failures followed by a success, we do not have to worry about calculating the number of ways to obtain a particular outcome. If p is the probability of a success on a given try, the probability distribution for G is

$$P(G = n) = (1 - p)^{n-1} p. \qquad (3.62)$$

This is an example of a *geometric probability distribution*. A portion of this distribution is shown in figure 3.3 for $p = 0.25$. This appears to be a much simpler distribution than the binomial, but don't be fooled. Looks can be deceiving.

Let's find the expectation of a geometric random variable so we can estimate how frequently the fringehead wins. The value of the geometric random variable is the number of attempts (n) required to get a success. Therefore, for each possible outcome value, n, we must multiply by the probability of obtaining that value and then sum all the products. For $n = 1$, $P = p$, for $n = 2$, $P = (1 - p)p$, for $n = 3$, $P = (1 - p)^2 p$, and so forth. It is a simple task until you realize that there is no reason to stop. Although the probability

---

[4] So called because each successive term is a constant times the previous term. In this case, every term is simply the prior term multiplied by $(1 - p)$.

becomes exceedingly small as n increases, there is no limit to how large n can be. Of course, in our particular example the fish would die from accumulated wasted time if n got very large, but, in principle, for this random variable n is unbounded. To calculate $\mu(G)$ we have to sum the products all the way to an n of infinity.

Obviously, we cannot continue to use our brute-force approach of calculating the contribution to the expectation for each possible outcome. We would never finish. We even have to question whether this geometric distribution is a probability distribution at all. If there are an infinite number of possible outcomes, can the sum of all their probabilities still equal 1? Let's see.

$$\sum_{n=1}^{\infty} P(G = n) = \sum_{n=1}^{\infty} (1 - p)^{n-1} p.$$

$$= p \sum_{n=1}^{\infty} (1 - p)^{n-1}$$

$$= p \sum_{n=0}^{\infty} (1 - p)^{n}. \tag{3.63}$$

To continue, we have to dredge up memories from our introductory calculus class about infinite sums. The sum of a variable x^n as n goes from 0 to ∞ converges to $1/(1 - x)$ if $|x| < 1$. If this result doesn't ring any bells, see box 3.1 for the proof. Since $0 < (1 - p) < 1$ for all relevant values of p, the sum $\sum_{n=0}^{\infty} (1 - p)^n$ converges to $1/p$. Thus,

$$\sum_{n=1}^{\infty} P(G = n) = p \left(\frac{1}{p} \right)$$

$$= 1. \tag{3.64}$$

Despite an infinite number of possible outcomes, the geometric distribution does not violate rule 2 (chapter 2).

Box 3.1 Infinite sums.

The secret to solving infinite sums, such as $\sum_{n=0}^{\infty} x^n$, is to find a way to get rid of the counter n. This typically involves rewriting the sum in a form that involves a ratio or an exponent where n disappears when we take the limit as $n \to \infty$. To begin, let's write out our sum explicitly.

$$\sum_{n=0}^{k} x^n = 1 + x + x^2 + \ldots + x^{k-1} + x^k. \tag{3.65}$$

We then multiply by $1 - x$:

$$(1 - x) \sum_{n=0}^{k} x^n = (1 - x) \cdot (1 + x + x^2 + \ldots + x^{k-1} + x^k)$$

$$= (1 + x + x^2 + \ldots + x^{k-1} + xk)$$

$$- (x + x^2 + \ldots + x^{k-1} + x^k + x^{k+1}). \tag{3.66}$$

Canceling like terms on the right-hand side leaves

$$(1 - x) \sum_{n=0}^{k} x^n = 1 - x^{k+1}, \tag{3.67}$$

from which we obtain a new expression for our original sum,

$$\sum_{n=0}^{k} x^n = \frac{1 - x^{k+1}}{1 - x}. \tag{3.68}$$

Now we need to take the sum in the limit as $k \to \infty$:

$$\sum_{n=0}^{\infty} x^n = \lim_{k \to \infty} \left(\frac{1 - x^{k+1}}{1 - x} \right). \tag{3.69}$$

If $|x| < 1$, then $x^{k+1} \to 0$ as $k \to \infty$. So,

$$\sum_{n=0}^{\infty} x^n = \frac{1}{1 - x} \tag{3.70}$$

as originally proposed.

Now let's return to calculating the average time the fish must wait for its first win. From our definition of the expectation (eq. 3.6),

$$\mu(G) = \sum_{n=1}^{\infty} n(1 - p)^{n-1} p. \tag{3.71}$$

Let $q = 1 - p$ and expand the sum:

$$\mu(G) = p + 2qp + 3q^2 p + 4q^3 p + \ldots + nq^{n-1} p. \tag{3.72}$$

Now we need our trick (see box 3.1). Subtract $q\mu(G) = qp + 2q^2 p + 3q^3 p + \ldots$ from each side.

$$\mu(G) - q\mu(G) = p + qp + q^2 p + q^3 p + \ldots q^n p$$

$$(1 - q)\mu(G) = \sum_{n=1}^{\infty} q^{n-1} p$$

$$p\mu(G) = \sum_{n=1}^{\infty} (1 - p)^{n-1} p. \tag{3.73}$$

The right-hand side of this equation should look familiar. This is the sum of probabilities in our geometric random variable, G (eq. 3.62), which we have already shown equals 1 (as all probability distributions must). Thus,

$$p\mu(G) = 1$$

$$\mu(G) = \frac{1}{p}, \tag{3.74}$$

a pleasantly elegant solution. The expected time to the first success is equal to 1 over the probability of success. Therefore, even though there are an infinite number of possible outcomes, there is a finite expectation of this random variable. Solving for the variance and standard deviation of a geometric random variable involves an even less intuitive set of tricks, but equally simple answers. We will save you the footwork and just provide the results:[5]

$$\sigma^2(G) = \frac{1 - p}{p^2}. \tag{3.75}$$

So,

$$\sigma(G) = \frac{\sqrt{1 - p}}{p}. \tag{3.76}$$

If we apply these results to our fringehead problem, where $p = 0.25$, the fish can expect to win once every 4 matches, with a standard deviation of approximately 3.5 matches. The variation around the expectation could be very important in this type of a situation, since a long string of failures (= time away from shelter) could lead to dire consequences. Suppose a fish dies if it doesn't get shelter within 2 days (= 8 matches). What is the chance that this sort of run of bad luck will occur? We will consider such questions about extreme events in detail in chapter 7, but we can use a brute-force approach to answer this specific question here. Returning to the probability distribution for the geometric random

[5] If you would like to see the derivations, consult Kinney (1997).

variable, we can calculate the probability of winning in anywhere from 1 to 8 matches. One minus the sum of these terms will then provide the probability of *not* winning within two days (= $P(Fish\ Dies)$).

$$P(1) = \frac{1}{4}$$

$$P(2) = \frac{3}{16}$$

$$P(3) = \frac{9}{64}$$

$$P(4) = \frac{27}{256}$$

$$P(5) = \frac{81}{1024}$$ \hfill (3.77)

$$P(6) = \frac{243}{4096}$$

$$P(7) = \frac{729}{16384}$$

$$P(8) = \frac{2187}{65536}.$$

The sum of these eight terms is approximately 0.90. This means the fish has about a 10% chance of dying before his next win.

3.8 *Summary*

In this chapter we expanded the simple rules of probability to address problems in which each outcome can be associated with a numerical response (= a *random variable*). We focused on discrete examples, postponing discussion of continuous random variables to the next chapter. Because random variables produce a number associated with each outcome, repeated experiments generate distributions of values with an average and variation around the average. We developed the notions of *expectations*, *variances*, and *standard deviations* to quantify the average and variation for any given random variable. Fortunately, there are a small number of probability distributions that can be used to describe a wide range of biological and physical phenomena. These distributions are also related to one another in relatively simple ways. We started with the notion of random variables (*Bernoulli variables*) that had only two possible values (success and failure). Next, we showed that combining several Bernoulli trials into a more complex variable generates a second useful distribution—the *binomial distribution*. Finally, we examined a special case of the binomial distribution where you have a string of failures followed by a single success—the *geometric distribution*. This last distribution forced us to deal with situations where there are an infinite number of possible outcomes. Surprisingly, the sum of the probabilities of these infinite numbers of outcomes summed to 1, and there was a finite expectation. We will face more challenging problems with infinite numbers of outcomes in the next chapter when we deal with continuous probability distributions.

3.9 *Problems*

1. In a simple game of chance, you roll a single die. Every time you roll a 6, $50 is added to your casino account, every time you don't roll a 6, $10 is subtracted from your account. At the end of ten rolls, you receive your earnings or pay your debt.

   a. In these ten rolls of the die, how much money would you expect to win or lose?

   b. In these ten rolls, how much is your account likely to fluctuate?

   c. If you start with $30 in your account, what is the probability that you will be broke or in debt after ten rolls of the die?

   d. The casino decides that it will no longer allow you to go temporarily in debt: as soon as your account is $0 or less you are broke. Under these circumstances, if you again start with $30 what is the probability of going broke in ten rolls of the die?

2. Question 1 is analogous to many situations in biology. For example, there is chance involved in pursuing prey. If you catch a meal, you receive a substantial net caloric input. If the prey escapes, you have expended calories. Compare your answers to questions 1.c and 1.d in this biological context.

3. In eq. (3.74) we learned that the expected value of a geometrical distribution is $1/p$, where p is the probability of success. Thus, if there is a 10% chance of success in any one trial, you expect to wait ten trials before your first success. Given this probability of success, on what trial are you most likely to have your first success? Discuss the difference between the expressions "most probable" and "expected" as used in statistics.

4. The population of Lesser Boobies on a hypothetical island is decreasing; there are only n adults left. For reasons that will become clear in a moment, n is an even number. As a means of preserving the species, you capture r adult birds at random ($r < n/2$) and bring them back to your aviary, assuming that they will breed in captivity. Only after capturing the birds do you find out that in the wild, all adult boobies are mated for life. As a result, your well-meaning (but under-researched) effort at conservation will succeed only if by chance you have captured at least one mating pair of boobies in your random sample. What is the probability that no mating pair was captured?

5. During World War II, in an attempt to save money on expensive blood tests, the army came up with the following plan. Instead of testing every soldier's blood individually for the presence of a disease, they combined the blood of a group of soldiers and tested the mixture. If the test results were negative, they knew all soldiers in the group did not have the disease. They did not have to waste money testing these soldiers individually. If the group test was positive, they had to test everyone in the group. Under what conditions will this testing strategy save money? That is, when will the expected number of tests conducted using groups be less than the number of tests required without grouping?

6. As a promotion, a fast-food restaurant is giving away collectible cards with pictures of the starting players on your local sports teams. Each time you go to the restaurant, you get a random card.

    a. If the cards are for basketball, how many times would someone have to visit the restaurant on average to get at least one card for each of the five starting players?

    b. If the cards are for baseball, how many times would someone have to visit the restaurant on average to get at least one card for each of the nine starting players?

4

Continuous Patterns of Disorder

Life is not always discrete. As a result, there are many circumstances in biology in which the outcome of an experiment does not readily correspond to an integer value. For example, when two lions fight for a prey item, the approach we have taken so far is to assume that one predator wins (and therefore is free to consume the entire prey), while the other loses and gets nothing. In reality, it is quite possible for each predator to make off with a fraction of the kill. Two arms and a leg might amount to 36% of the overall calories in a baboon, and this might be the take-home result for one lion, the remaining 64% going to her competitor. But we cannot ask the binomial distribution to tell us what the probability is of getting 36% of a success; the binomial distribution pertains only to discrete, integer numbers of success and failure. In cases such as this, we must deal with continuous probability distributions in which the degree of success can take any of an infinite number of values. These distributions are the subject of this chapter. We will begin by examining the simplest form of a continuous distribution, a uniform distribution, and then tackle the workhorse of statistics, the normal probability density function.

The most fundamental change from discrete to continuous sample spaces is in how we deal with events. For discrete sample spaces, an event is a collection of outcomes in the sample space. Each outcome has some probability of occurrence, and the probability of the event is the sum of the probabilities of the outcomes in the event. This leads to a basic problem for continuous sample spaces. Recall that, by definition, the probability of an outcome is the number of times in which that outcome occurs divided by the total number of outcomes. But as we noted above, the total number of outcomes in a continuous sample space is infinite. There are a few special cases in which a sample space contains an infinite number of outcomes, yet each outcome has finite probability (recall the geometric distribution from the last chapter). But this is not the case with a continuous sample space. A continuous sample space has an uncountably infinite number of outcomes, and as a result the probability of any particular outcome is exactly zero. An infinite collection of outcomes that have no probability of occurring is a slippery notion. How can we sum the probability of individual outcomes to calculate the probability of events when each outcome has zero probability? In this respect, thinking about the probability of events in a continuous sample space is akin to thinking about paint on a wall. There is an infinite number of geometrical points on the wall. As a result, it takes zero paint to cover each point. Nonetheless, it takes a finite volume of paint to cover the wall. How can we reconcile these facts?

To deal effectively with continuous sample spaces, we have to shift our thinking from specific outcomes to *intervals* of outcomes. As we noted above, a point on the wall takes up no space, but we can easily measure how much paint it takes to cover a given area. Similarly, in a continuous sample space a specific outcome has zero probability of occurring, but an interval on our continuous scale (e.g., a number between 0 and 1, or a time between 12:00 and 12:15) *can* have a finite probability. In essence, intervals break up a continuous scale into a set of discrete chunks, and we use these discrete chunks the way we previously used discrete outcomes. Thus, instead of thinking of events such as $x = 2$, we focus on events such as $\{1 < x < 2\}$, and ask what the probability is that a random variable will lie within this range of values instead of at some particular point.

4.1 *The Uniform Distribution*

Let's first consider a simple probability distribution called the *uniform distribution*. Now, a discrete uniform distribution is one where all outcomes within a specified range have an equal probability, and the sum of the probabilities of the outcomes is 1. For example, in figures 4.1A and B the uniform probability distributions are plotted for two and ten outcomes. Because each distribution is discrete, the probability of any outcome is $1/n$, where n is the total number of outcomes in the sample space.

By analogy, a continuous uniform distribution is one where any *interval* within a given range has the same probability as any other *interval* of the same size, and the sum of the probabilities of all intervals is 1. But we face a problem when we try to take the next step and graph our distribution. We can divide the overall range of the sample space into n intervals of equal size, each with a probability of $1/n$. The larger n is, the smaller each interval becomes, and the smaller the probability associated with it. For a discrete distribution, n is set by the number of outcomes, and as long as n is finite there is no problem. But for a continuous distribution, the number of outcomes is uncountably infinite. Taken to the limit, the end result is the totally uninformative graph shown in figure 4.1C. This problem sets the stage for two alternative representations of a continuous probability distribution, one building on the other.

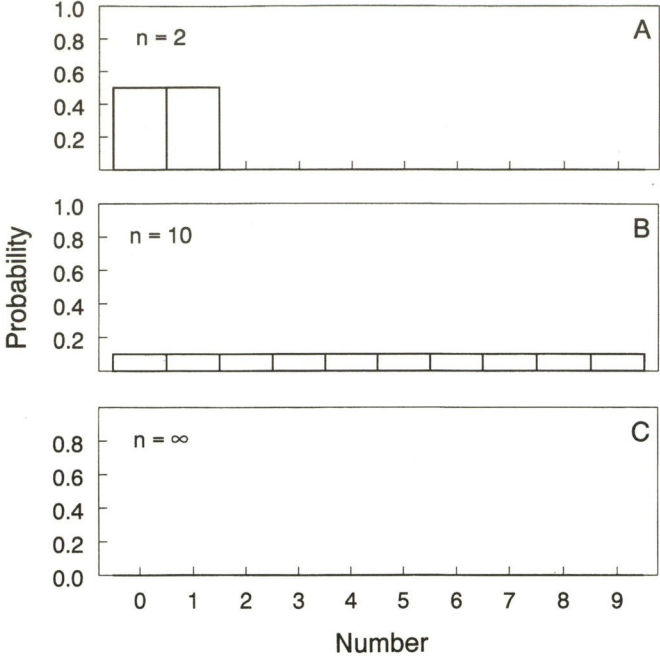

FIG. 4.1 Examples of discrete, uniform probability distributions for different numbers of outcomes. As n, the number of outcomes, increases from 2 (A) to 10 (B) to infinity (C), the distribution becomes flatter and flatter until it is indistinguishable from the x-axis.

4.1.1 THE CUMULATIVE PROBABILITY DISTRIBUTION

First let's explore something called a *cumulative probability distribution*. Consider a continuous uniform distribution with a range between $x = a$ and $x = b$. We divide this range into n intervals, each of width $(b - a)/n$. Because the distribution is uniform and intervals have the same size, each interval has probability $1/n$. As a result, the probability that x is in the lowest interval $(P(a \leq x \leq a + (b - a)/n))$ is $1/n$. The probability of having x in either this first interval or the one next to it (that is, $P(a \leq x \leq a + 2(b - a)/n)$) is $2/n$. More generally, the probability that x lies in the first i intervals $(a \leq x \leq a + i(b - a)/n)$ is i/n. Thus, as x extends from a to b, i goes from 0 to n, and the probability $P(x \leq x_i)$ increases from 0 to 1. In other words, as x goes from a to b we accumulate probability at a steady rate.

A plot of this accumulated probability as a function of x is shown in figure 4.2. Note that we get the same graph even as n gets infinitely large; its shape is independent of the number of intervals chosen. This type of graph is

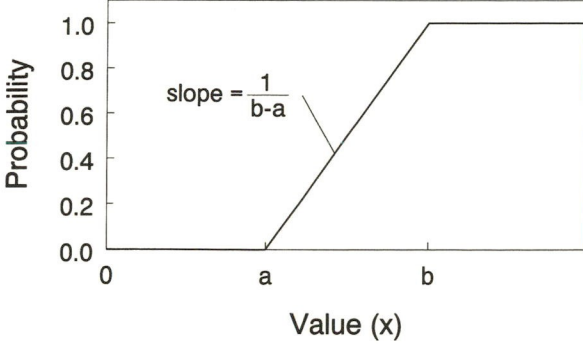

FIG. 4.2 The cumulative probability distribution for a stochastic process with a continuous, uniform chance of occurring. The values a and b bound the range in which events can occur, and the sloped line depicts the probability that a random event will have less than a specified value. In effect, as x moves from a to b, probability accumulates at a steady rate.

a cumulative probability distribution. Each point on the graph is $P(X \leq x)$, the probability that the random variable X is less than or equal to a particular value, x. We will find the cumulative probability distribution to be a useful tool.

4.1.2 THE PROBABILITY DENSITY FUNCTION

Although the cumulative probability distribution gives us a practical way of viewing the probability of a continuous random variable, there is an alternative that is equally useful. For many questions, we are not interested in finding the probability of an outcome being less than a particular value. Rather, we ask, what intervals of our random variable are most likely to occur?

To answer this question, we have to turn to the *slope* of the cumulative probability distribution curve. Intervals that contain values with a high probability of occurrence are characterized by a rapid rise in the cumulative probability distribution (i.e., a steep slope). In contrast, intervals containing values with a low probability of occurrence are characterized by only a gradual increase of the cumulative probability distribution (i.e., a slope near zero). Thus, the slope of the cumulative probability distribution measures how rapidly probability is accumulating at any particular point within the range of possible values, and is therefore a measure of how probability is "concentrated" along the axis of possible values of the random variable. Plotting the slope of the cumulative probability distribution gives us a picture of how probability is distributed along the axis for a continuous variable. Such a plot is analogous to the discrete probability distributions we plotted earlier. In probability theory, the equation describing the slope of the cumulative probability distribution is called the *probability density function* (often abbreviated as *pdf*).

DEFINITION 5 *If $G(X) = P(X < x)$ is the cumulative probability distribution of the random variable X, the probability density function, $g(X)$, is defined as*

$$g(X) \equiv \frac{dG(X)}{dx}.$$

The probability density function for the continuous uniform distribution is plotted in figure 4.3. Since the cumulative probability distribution increases linearly for this random variable (see fig. 4.2), the slope is constant within the range of possible values. Because the cumulative probability increases steadily from 0 to 1 over a range of length $(b - a)$, the slope is $1/(b - a)$, and this is the probability density function.

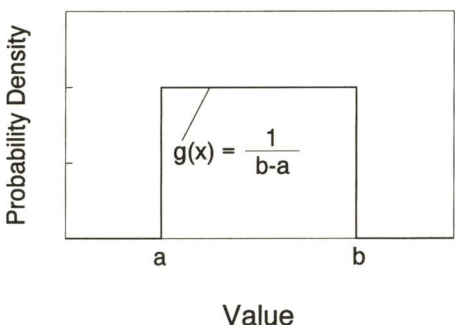

FIG. 4.3 The probability density function, $g(X)$, for a stochastic process with a continuous, uniform chance of occurring within the range bounded by a and b. Within this range, the value of the probability density function is equal to the slope of the line in figure 4.2. Outside of the specified range, $g(X) = 0$.

Remember, the probability density function does not measure the probability of getting a specific value x, because we know that the probability of any particular value is zero. Rather, the probability density function measures how rapidly the cumulative probability increases in the vicinity of x.

One nice feature of the probability density function is that the integral of the function over all possible values is always 1 (that is, the area under the whole probability density curve = 1). This is a result of the fundamental rule of integral calculus. The probability density function is the derivative of the cumulative probability. Integrating the derivative therefore yields the amount of probability accumulated, and the total probability of an outcome falling within the overall range of possible values has to be 1. Put mathematically:

$$\int_{x_{\min}}^{x_{\max}} g(x)dx = 1, \tag{4.1}$$

where the limits to integration (x_{min} and x_{max}) are the lower and upper bounds of the possible range of values.

This equality provides an alternative means to calculate the probability density function for a uniform, continuous probability distribution. We start with eq. (4.1) with lower and upper bounds of a and b, respectively:

$$\int_a^b g(x)dx = 1. \tag{4.2}$$

Knowing that for the uniform distribution $g(x)$ is constant allows us to bring it outside the integral:

$$g(X) \int_a^b dx = 1. \tag{4.3}$$

From the definition of an integral we then obtain

$$g(X) \cdot x|_a^b = 1,$$

$$g(X)(b - a) = 1,$$

$$g(X) = \frac{1}{(b - a)}, \tag{4.4}$$

the conclusion we arrived at earlier.

Using the same logic, it follows that the area under any interval of the probability density function gives us the probability of the random variable occurring within that interval:

$$\int_c^d g(x)dx = P(c \le x \le d). \tag{4.5}$$

As a result, if you know the probability density function for a random variable, you can calculate the probability that an outcome falls within any interval of values. Consider, for instance, the sort of spinning arrow that is included with many board games. If you give this arrow a spin, it has equal probability of stopping at any angle x between $0°$ and $360°$. As a result, we know that $g(X) = 1/360$. The probability that the arrow will stop between $5°$ and $13°$ is

$$\int_5^{13} g(x)dx = \int_5^{13} \left(\frac{1}{360}\right)dx$$

$$= \frac{1}{360} \int_5^{13} dx$$

$$= \frac{1}{360} \cdot (13 - 5)$$

$$= \frac{8}{360}. \tag{4.6}$$

In this example, $g(X)$ is a constant because we are dealing with a uniform distribution, and this simplifies the evaluation of the integral. *But the basic idea is valid whether $g(X)$ is a constant or not.* For example, consider the probability density function shown in figure 4.4. This probability density function varies with x, but the probability that x lies between a and b is nonetheless equal to the cross-hatched area under the curve. We will explore several such cases in which the value of the probability density function varies with x.

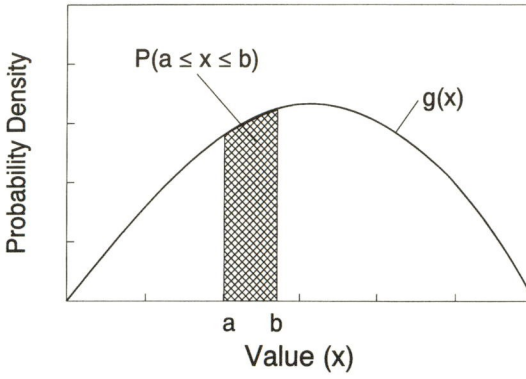

FIG. 4.4 The probability that x will occur within the range bounded by a and b is equal to the area under the probability density function within this range (see eq. 4.5).

4.1.3 The Expectation

Now that we have a means of characterizing continuous random variables, it will be useful to determine their average and variance. Recall that for a discrete random variable, the average value was calculated by summing the products of each possible outcome and its probability of occurrence:

$$E(X) = \sum_{\text{all } x} x P(x). \tag{4.7}$$

But if we apply this same logic to continuous variables, we have a problem. As we have discussed, there is an infinite number of outcomes, and any specific outcome has zero probability. Summing this particular infinite series would be problematic. Once again, we solve the problem by shifting our attention from outcomes to intervals.

Let's divide up the range of possible values into many intervals of constant width and treat each interval in the same fashion as we treated outcomes in eq. (4.7). This process is shown in figure 4.5 for one particular interval size. To carry out this calculation, we need a value for the outcome associated with each interval and the probability that an outcome will lie within this interval.

First, the value. As long as we divide our overall range into a large number of intervals, the width of each interval, Δx, is relatively small, and we can accurately characterize the values in that interval by using the interval's midpoint, x_m.

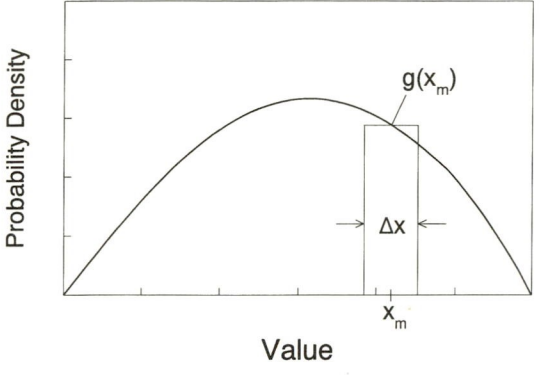

FIG. 4.5 We can estimate the area under a portion of the probability density curve using a rectangle as shown here. The width of the rectangle is Δx (corresponding to a small interval of x), and its height is equal to the value of the probability density function at x_m, the midpoint of the interval. The smaller Δx is, the more accurate our estimate is.

Next, we need a formula for the probability of an outcome occurring within each interval. As we noted above, the probability that an outcome occurs within any given interval is equal to the area under the probability density function bounded by the interval. If we again assume that intervals are narrow, we can approximate the area under the probability density function by calculating the area within a rectangle of width Δx and height $g(x_m)$, where x_m is again the midpoint of the interval. Our estimate (call it Δp) of the probability that an event occurs within a particular interval is thus

$$\Delta p = g(x_m) \cdot \Delta x. \tag{4.8}$$

We now have both components that we need to estimate the expected value using the discrete random variable approach:

$$E(X) = \sum_{\text{all intervals}} [x_m \cdot g(x_m)\Delta x]. \tag{4.9}$$

That is, we take the value associated with each interval (x_m), multiply by the probability associated with each interval, $g(x_m)\Delta x$, and sum the product over all intervals.

The accuracy of this approximation depends on the width of the intervals. If they are wide, the use of the midpoint value may not be an accurate approximation for either x or $g(X)$ across the whole interval. Conversely, we can increase the accuracy of the estimate by making the interval widths as small as possible. Mathematically we do this by evaluating the formula in the limit as Δx approaches 0.

$$E(X) = \lim_{\Delta x \to 0} \sum_{\text{all } x} [x_m \cdot g(x_m)\Delta x]. \tag{4.10}$$

If you remember your introductory calculus, this reduces to the following definition:

DEFINITION 6 *The* expectation *of a continuous random variable, X, is*

$$E(X) = \mu_X \equiv \int_{-\infty}^{\infty} xg(x)dx.$$

In other words, the expectation of a continuous random variable is the integral of the product of possible outcomes and the value of the probability density function associated with each outcome. Here, the probability density function is the weighting that is analogous to the probability in the discrete case.

Let's try this formula on the uniform random variable in figure 4.3. The random variable is constrained to lie between a and b, and the probability density function is constant $(= 1/(b-a))$ within this range. Therefore,

$$\mu_X = \int_a^b \left(x \cdot \frac{1}{(b-a)} \right) dx$$

$$= \frac{1}{(b-a)} \cdot \frac{x^2}{2} \Big|_a^b$$

$$= \frac{b^2 - a^2}{(b-a) \cdot 2}$$

$$= \frac{b+a}{2}. \qquad (4.11)$$

Thus, the expected value is the midpoint of the range of possible outcomes, a conclusion that matches our intuition.

4.1.4 THE VARIANCE

The variance of a continuous random variable is defined by a similar analogy to the discrete case:

DEFINITION 7 *The* variance *of a continuous random variable, X, is*

$$E\big[(X - \mu_X)^2\big] = \sigma_X^2 \equiv \int_{-\infty}^{\infty} (x - \mu_X)^2 g(x)dx.$$

For example, the variance of a random variable with a uniform probability density function with a range between a and b is

$$\sigma_X^2 = \int_a^b (x - \mu_X)^2 g(x)dx$$

$$= \int_a^b \left(x - \frac{b+a}{2} \right)^2 \frac{1}{b-a} dx. \qquad (4.12)$$

When the dust settles on the algebra, this turns out to be

$$\sigma_X^2 = \frac{(b-a)^2}{12}.$$ (4.13)

Taking the square root, we see that the standard deviation of a uniform distribution is

$$\sigma_X = \frac{|b-a|}{\sqrt{12}} \cong \frac{|a-b|}{3.46}.$$ (4.14)

Now $1/3.46 = 0.289$, so the standard deviation of a uniform distribution is approximately 29% of the overall range of the variable.

4.2 *The Shape of Distributions*

At this point let us digress to consider the *shape* of probability distributions. The ideas developed here will become important in a moment when we explore the normal distribution.

First, consider the simple case of a discrete, uniform probability distribution of the variable X with a range equal to n, the number of outcomes. If there are two outcomes, the range extends from $x = 0$ to $x = 2$, and each outcome has probability 1/2. If there are three outcomes, the range is from 0 to 3, and each outcome has probability 1/3. In general, if there are n outcomes, the range is from 0 to n, and each outcome has probability $1/n$. A series of these distributions is shown in figure 4.6. Outcomes are spaced equally along the x-axis.

Note that the "location" and shape of these distributions changes as n increases. By "location" we mean the center of the distribution, its average, μ_X (shown here by the arrows). From the definition of the mean, we know that $\mu_X = n/2$, so μ_X moves farther and farther out on the x-axis as n increases. In addition, the distribution becomes flatter and flatter as n increases because the probability is spread among a larger number of outcomes. In other words, the distribution's shape changes as a function of n.

Now for some fun. For each of the distributions shown in figure 4.6 let's try the following procedure. First, let's relocate the distribution so that it is centered around 0. We can do this by subtracting μ_X from each value of x. The shifted distribution now extends from $-n/2$ to $n/2$.

Next, we adjust the shape of the distribution in two steps: we divide each shifted value of x by the magnitude of the overall range (n), and we multiply each probability value by n. Dividing each x by n results in a transformed range that now extends from $-1/2$ to $1/2$. In each distribution, the probability of each x was $1/n$, so that when we multiply by n, each transformed probability

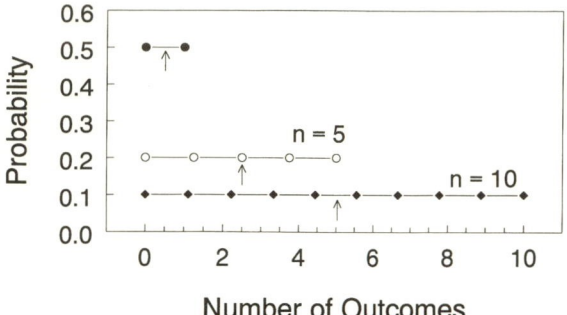

FIG. 4.6 Discrete, uniform probability distributions are shown for 2, 5, or 10 outcomes. To preserve visual clarity, the probability associated with each outcome is shown as a dot rather than the traditional histogram bar, but the idea is the same. As the number of outcomes increases, the mean value of the distribution (shown by the arrows) moves to larger values, and the probability associated with any particular outcome decreases.

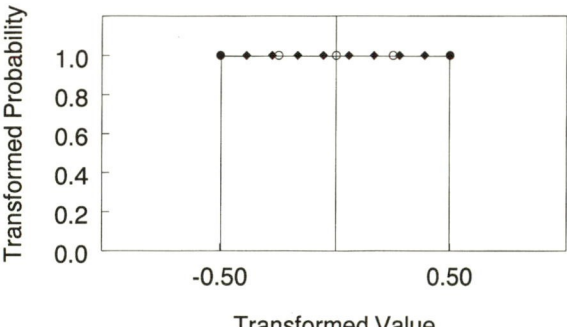

FIG. 4.7 The discrete, uniform probability distributions of figure 4.6 have been transformed as described in the text. The result is a distribution that has the same shape independent of the number of outcomes. As the number of outcomes increases, this transformed distribution begins to look like the probability density function shown in figure 4.3. The symbols here correspond to those used in figure 4.6.

is now equal to 1. As a result of this massaging, the *shape* and *location* of each transformed distribution is now exactly the same (fig. 4.7). In particular, this constancy of shape holds even as *n* increases without bound.

So what? Let's think for a moment about what we have done in applying this transformation. Each of our original *x* values (each signifying a particular outcome) was spaced a distance of $\Delta x = n/(n-1)$ from its neighbors, but in the transformed version their spacing is now $1/(n-1)$. In the limit as n increases toward infinity, the spacing between points in our transformed distribution gets smaller and smaller, and we in spirit move from a discrete distribution of points to a continuous distribution extending across our transformed range. We know that as n increases, the probability associated with each of these points decreases, and as n approaches infinity the probability approaches zero.

But this transformation of the x-axis from a discrete distribution to something approaching a continuous one is only half of how we have massaged the distribution. Recall that we simultaneously multiplied each probability value by n. By doing so, we have created a new "variable" related to the probability, but this variable stays constant at 1 as n increases. What exactly is this new variable? We can gain a hint by calculating the area bounded by the graph of our transformed distribution (see fig. 4.7). This is always a rectangle with a base of $1/2 - (-1/2) = 1$ and a height of 1. Thus, no matter which of our uniform distributions we transform, the area within the transformed shape is 1. In effect, our transformation has converted a discrete probability distribution (in which the sum of all probabilities is 1) to a *probability density function* (in which the area under the curve is 1). Note, however, that this standardized probability density function embodies the essential characteristic of the discrete uniform probability distribution from which it is transformed: probability is still spread uniformly across the range of outcomes.

4.3 *The Normal Curve*

Now let's try the same type of transformation with a different distribution. Figure 4.8 shows a series of binomial distributions with $p = 0.5$ and increasing n. Much like the uniform distributions above, these binomial distributions change their location and shape as n increases. The center of the distribution (its mean) moves to higher and higher values, and the height of the distribution decreases as probability is spread among more outcomes.

What happens if we now transform these binomial distributions? As before, we adjust the location by subtracting the mean, but in this case, we use a slightly different procedure to adjust the shape. Rather than dividing each x value by the total range of the distribution, we instead divide by an alternative measure of the distribution's spread, the standard deviation, σ_X. This transformed variable, $(x - \mu_X)/\sigma_X$, is given a new name, Z. As for the values of probability, rather than multiplying each probability value by n, in this case we multiply by the standard deviation.

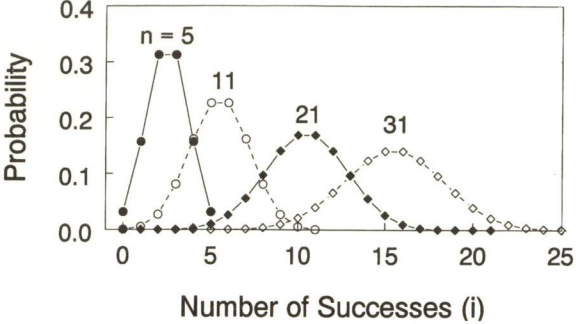

FIG. 4.8 Binomial probability distributions are shown for 5, 11, 21, and 31 trials. As the number of trials (n) increases, the mean of the distribution (its peak) moves to larger values, and the probability decreases that any particular number of successes will be obtained.

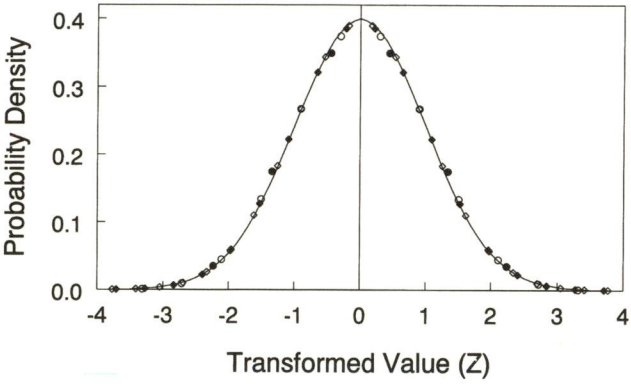

FIG. 4.9 The binomial distributions of figure 4.8 have been transformed as described in the text. The result is a "bell-shaped" distribution that has the same shape independent of the number of outcomes. As the number of outcomes increases, this transformed distribution asymptotes to the standard normal probability density function (the solid line, eq. 4.15). The symbols here correspond to those used in figure 4.8.

The results are shown in figure 4.9. The effect is again to maintain the shape and location of the transformed distribution as n increases without bound. For large n, our transformed, discrete distribution approximates a continuous distribution (in that the outcomes become more closely spaced and the probability associated with each point on the x-axis approaches 0), but because we have multiplied each probability value by σ_X, the transformed value remains finite.

In fact, the area under the transformed curve is 1! In other words, by applying this particular transform to the binomial distribution and allowing n to become large, we arrive at a probability density function with the distinctive "bell" shape shown in figure 4.9. It is a symmetrical distribution that peaks in the center and falls rather rapidly to low values as you move in either direction from the midpoint. In that it expresses how probability varies across the range of outcomes, this probability density function retains the essential character of the binomial distribution from which it is derived.

This peculiar bell-shaped curve was discovered in the 1700s by Jacob Bernoulli and Abraham De Moivre, and it is called the *standard normal probability density function*. Note that because of the manner in which the binomial distribution is transformed to arrive at the standard normal curve, the resulting probability density function has a mean of 0 and a variance and standard deviation both equal to 1. Bernoulli and De Moivre were able to show that the shape of this function is described by the following equation (see appendix 1 to this chapter):

$$g(Z) = \frac{1}{\sqrt{2\pi}} \exp\left(-\frac{z^2}{2}\right). \tag{4.15}$$

Here "exp" is the exponential function: $\exp(x) = e^x$. The discovery of the standard normal probability density function has had an enormous impact on probability theory and statistics; you will see it (or a variation on it) reappear many times in the coming chapters.

Now, the probability density function described by eq. (4.15) is a distillation of a whole family of probability density functions that one might encounter in the real world. It has the characteristic shape of each of these real-world functions, but is constrained to have $\mu_X = 0$ and $\sigma_X = 1$. We can, however, convert the standard normal curve to describe similar curves that have other means and standard deviations. To convert from the variable Z used in the standard normal, to the variable X used in a particular experiment, we essentially untransform our variable by employing the following formula:

$$X = (\sigma_X \cdot Z) + \mu_X. \tag{4.16}$$

Thus, multiplying each Z value by the particular standard deviation associated with an experiment (σ_X) and adding the experiment's mean (μ_X), yields a real-world variable whose probability density function has the shape embodied by the standard normal curve. The mean and standard deviation of this new curve are μ_X and σ_X.

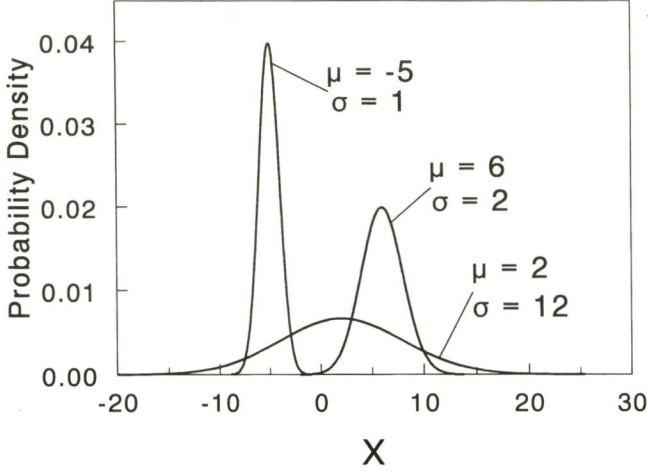

Fig. 4.10 Examples of the normal probability density function. By varying the mean, μ_X, and the standard deviation, σ_X, the location and width of the curve can be adjusted.

With some mathematical machination, De Moivre and Bernoulli showed that the probability density function for this new variable is

$$g(X \mid \mu_X, \sigma_X) = \frac{1}{\sigma_X \sqrt{2\pi}} \exp\left[-\frac{1}{2} \frac{(x - \mu_X)^2}{\sigma_X^2}\right]. \tag{4.17}$$

As we noted above, by varying σ_X and μ_X, one can create a whole family of distributions, collectively called *normal* or *Gaussian distributions*.[1] A variety of normal curves are shown in figure 4.10.

4.4 *Why Is the Normal Curve Normal?*

The primary significance of the normal distribution is that many chance phenomena are at least approximately described by a member of the family of normal probability density functions. If you were to collect a thousand snowflakes and weigh each one, you would find that the distribution of their weights was accurately described by a normal curve. If you measured the strength of bones in wildebeests, again you are likely to find that they are normally distributed.

[1] Named in honor of Johann Friederich Carl Gauss, one of the great minds in mathematics. Although the bulk of Gauss's work dealt with number theory, he served as a consultant on a project designed to measure the curvature of the earth. In the process of coping with the inevitable errors made in these real-world measurements, Gauss made fruitful use of the normal curve (Bernstein 1996).

Why should this be so? Recall that the standard normal distribution was derived (in essence) by summing many independent Bernoulli random variables to give a binomial distribution with large n. It turns out that if we add together many random variables, all having the same probability distribution, the sum (a new variable) has a distribution that is approximately normal. The amazing fact is that this normality *does not depend on the probability distribution of the individual random variable in the sum*! In other words, *taking the sum of a large number of almost any kind of random variables creates a new random variable that is approximately normal*. This result is formally called the *Central Limit Theorem*, and it provides the theoretical basis for why so many variables that we see in nature appear to have a probability density function that approximates a bell-shaped curve.[2] If we think about random biological or physical processes, they can often be viewed as being affected by a large number of random processes with individually small effects. The sum of all these random components creates a random variable that converges on a normal distribution regardless of the underlying distribution of processes causing the small effects.

For future reference, we note a common situation to which the Central Limit Theorem applies. Imagine that you need to know the size of barnacles. To this end, you go to the shore, select a random sample of barnacles, and measure the diameter of each individual. They aren't all the same size, so you average the measured values and use this sample mean as a representative estimate of size. As a check on this value, you choose another random sample, measure them, and calculate a second sample mean. Because there is variation among barnacles and you are picking them at random, it is unlikely that this second sample mean exactly equals the first. What can we say about the variation among sample means? This question is easily answered when we realize that a mean is the result of summing a number of independent random variables (in this case, the size of individual barnacles). As a consequence, we can use the the Central Limit Theorem to assert that the means of our samples of barnacle size (as with the means of any independent samples of a given population) will be normally distributed. Much use is made of this fact in inferential statistics.

The Central Limit Theorem is a powerful tool, and we wish we had an intuitive explanation of why it should be true. Unfortunately, we don't. The best we can do is provide an informal demonstration of it validity, which, if you are feeling adventuresome (or skeptical) can be found in appendix 2 at the end of this chapter.

---

[2] The central limit theorem was first proposed by Simon Laplace in 1809.

4.5 *The Cumulative Normal Curve*

Although the normal probability density function has wide applicability, there are occasions where its use is enhanced if it is converted to a cumulative probability distribution. For example, if we know that the height of adult males in our town is normally distributed with a mean of 70 inches and a standard deviation of 3 inches, we might be interested in the question: What is the probability that the next man we meet will be shorter than 61 inches? This is the type of question that the cumulative probability curve can answer directly:

$$P(x < 61) = \int_0^{61} g(x)dx$$

$$= \int_0^{61} \frac{1}{\sigma_X\sqrt{2\pi}} \exp\left[-\frac{1}{2}\left(\frac{x - \mu_X}{\sigma_X}\right)^2\right]dx, \qquad (4.18)$$

where $\mu_X = 70$ and $\sigma_X = 3$. All we need to do is evaluate this integral. There is one slight problem, however. The form of this integral is such that it cannot be solved analytically. In other words, we cannot write an equation in terms of x, σ_X, and μ_X that will directly provide the answer. Instead, we have to rely on estimates of the answer that have been tabulated through an iterative numerical process. That is, we have to look it up.

This would be a real problem if we had to have a separate table of results for each combination of x, σ_X, and μ_X. It is here that the distribution of Z really shines. Because the standard normal distribution describes the shape of any normal distribution, all we really need is a table that gives us the proportions of the standard normal distribution. We can then take the problem at hand (which sets the values of x, σ_X, and μ_X) and normalize it by subtracting μ_X from x and dividing the result by σ_X. This result can then be looked up in a Z table (an abbreviated version is given in table 4.1, a more complete table can be found in most introductory statistics texts). Note that the table only covers positive values of Z. Utilizing the symmetry of the normal curve, we note that $P(-Z) = 1 - P(Z)$, and this equality can be used to calculate probabilities corresponding to negative values of Z. For example, $P(1.3) = 0.9032$. Thus, $P(-1.3) = 1 - 0.9032 = 0.0968$. The cumulative standard normal probability distribution can also be shown graphically (fig. 4.11).

We can use this table to answer the question posed above: What is the probability that the next man we encounter is shorter than 61 inches? First we calculate the appropriate Z value by subtracting 70 (the mean) from 61 and dividing the result by 3 (the standard deviation). $Z = -3$. In other words, a man 61 inches tall would be three standard deviations below the mean. A quick calculation using table 4.1 tells us that the probability of meeting a man this short or shorter would be 0.0014.

TABLE 4.1 An Abbreviated Table of the Dimensions of a Standard Normal Curve

Z	0.0	0.1	0.2	0.3	0.4	0.5	0.6	0.7	0.8	0.9
0	0.5000	0.5398	0.5793	0.6179	0.6554	0.6915	0.7257	0.7580	0.7881	0.8159
1	0.8413	0.8643	0.8849	0.9032	0.9192	0.9332	0.9452	0.9554	0.9641	0.9713
2	0.9773	0.9821	0.9861	0.9893	0.9918	0.9938	0.9953	0.9965	0.9974	0.9981
3	0.9986	0.9990	0.9993	0.9995	0.9997	0.9998	0.9998	0.9999	0.9999	1.0000

Notes: The value of Z is given by the numbers in the left-most column and the column headings. For example, if you run your finger down to the row labeled "1" and move across to the row labeled "0.3," you find the value 0.9032. This is the cumulative probability associated with $Z = 1.3$.

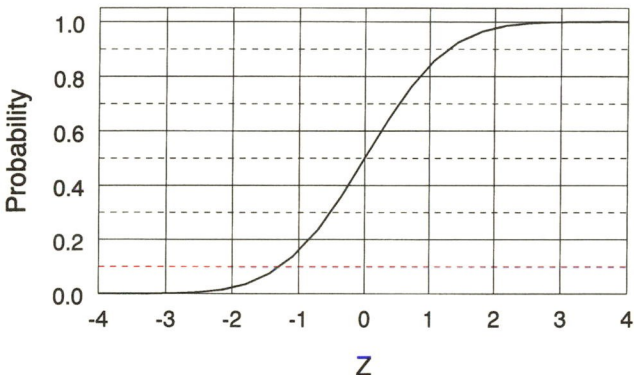

FIG. 4.11 The cumulative standard normal probability distribution. The values shown in this graph are tabulated in table 4.1.

You may have noticed that the use of the normal distribution in this fashion requires a set of information that is different from the one we previously used when dealing with the binomial distribution. For example, in the question we have just examined, we were given the mean and standard deviation of the population, and from these we calculated a cumulative probability. In contrast, when using the binomial distribution, we were given the number of trials (n), probabilities of success and failure (p and q, respectively), and the values associated with success and failure (V and W, respectively), and from these data we were asked to calculate the appropriate probabilities. In fact, the two approaches are not as different as they might seem. Given n, p, q, V, and W, we can calculate both the mean ($np|V - W|$) and the standard deviation ($\sqrt{npq}|V - W|$) for the binomial distribution. These parameters can then be used with the formula for the standard normal curve (which, after all, is an approximation of the binomial distribution) to estimate the appropriate probabilities.

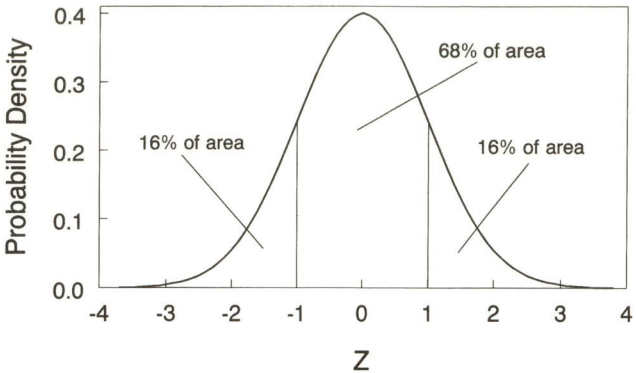

FIG. 4.12 Useful dimensions of the standard normal probability density function. There is a 68% chance that a random event will have a value within one standard deviation of the mean, that is, within the range $-1 < Z < 1$.

Note for future reference that in a normal distribution, 16% of all outcomes have a Z value less than -1, and 16% have a Z value greater than 1 (fig. 4.12). Thus, 32% of all outcomes are expected to lie more than one standard deviation away from the mean. Conversely, $1 - 0.32 = 68\%$ of all outcomes are likely to fall *within* a standard deviation of the mean (either above or below it). Similar calculations show that about 95% of all outcomes lie within two standard deviations of the mean, and about 99.9% lie within three standard deviations of the mean.

4.6 *The Standard Error*

Before we leave the subject of probability distributions and move on to their biological use, we must cover one final, important topic. Once again we return to our workhorse, the example of sarcastic fringeheads fighting for shelters. So far, we have calculated the expected daily time wasted due to these random encounters (the mean, μ_X) and a measure of the day-to-day variability in wasted time (the standard deviation, σ_X). At least one major question remains. We know that the expected daily time wasted, μ_X, is an idealized value based on the assumption of an infinite number of wrestling matches, and we can therefore guess that the empirically averaged time, $\overline{X}$ (measured over a month, say), will be different from the expectation. How different are these two values likely to be? To answer this question, we need to calculate a quantity known as the *standard error*.

DEFINITION 8 *The standard error, $\sigma_{\overline{X}}$, is the standard deviation of the empirical mean from the expectation.*

This is easily stated, but how do we calculate the standard error? Once again we return to our wrestling fringeheads.

After a month of interacting with his peers, a fish has amassed a total of thirty "samples" of the time wasted in a day, and by summing these and dividing by thirty we can arrive at a calculated average for the daily time wasted. Put mathematically, we calculate:

$$\overline{X} = \frac{1}{N} \sum_{i=1}^{N} x_i, \tag{4.19}$$

where $\overline{X}$ is the empirical average (as opposed to the idealized average, μ_X), x_i is the time wasted in a particular day, and (in this case) $N = 30$. As we have seen (chapter 3), if we were to repeat our experiment a very large number of times, this calculated average would approach the expectation, μ_X, as required by the Law of Large Numbers.

At this point, we have an estimate of the average daily time wasted based on a month's worth of data. By how much is our estimate likely to deviate from the expected time wasted? We are now in a position to make a calculation. Following the logic we used before in calculating a variance, we can calculate the expectation of the squared deviation of the empirical average $\overline{X}$ from the "true" average, μ_X:

$$
\begin{aligned}
E\left[\left(\overline{X} - \mu_X\right)^2\right] = \sigma_{\overline{X}}^2 &= E\left\{\left[\left(\frac{1}{N}\sum_{i=1}^{N} x_i\right) - \mu_X\right]^2\right\} \\
&= E\left\{\left[\left(\frac{1}{N}\sum_{i=1}^{N} x_i\right) - \left(\frac{1}{N}\sum_{i=1}^{N}\mu_X\right)\right]^2\right\} \\
&= E\left\{\frac{1}{N^2}\left[\left(\sum_{i=1}^{N} x_i\right) - \left(\sum_{i=1}^{N}\mu_X\right)\right]^2\right\} \\
&= E\left\{\frac{1}{N^2}\left[\sum_{i=1}^{N}\left(x_i - \mu_X\right)\right]^2\right\} \\
&= \frac{1}{N^2}E\left\{\left[\sum_{i=1}^{N}\left(x_i - \mu_X\right)\right]^2\right\}. \tag{4.20}
\end{aligned}
$$

At first glance, this last expression might appear to be complicated by the fact that we are calculating the expectation of a squared summation. Indeed, squaring the sum leads to a whole host of cross-products between x and μ_X. But, just as we found when we calculated the variance of the sum of variables, because the x's here are independent, the expectation of these cross-products is just zero; for independent values, the square of the sum is equal to the sum of the squares.

Working through the algebra, we find that

$$E[(\overline{X} - \mu_x)^2] = \frac{1}{N^2} E\left[\sum_{i=1}^{N}(x_i - \mu_x)^2\right]. \tag{4.21}$$

But

$$E\left[\sum_{i=1}^{N}(x_i - \mu_x)^2\right] = N\sigma_X^2, \tag{4.22}$$

(the expectation of a sum is the sum of the expectations). Thus,

$$E\left[(\overline{X} - \mu_x)^2\right] = \sigma_{\overline{X}}^2 = \frac{N\sigma_X^2}{N^2} = \frac{\sigma_X^2}{N}. \tag{4.23}$$

This is an important result. What it tells us is that the variance associated with our empirical estimate of the mean is equal to the variance of the distribution from which our samples have been taken, σ_X^2, divided by the number of samples that went into calculating $\overline{X}$. For the case at hand, we know that the variance of the daily time wasted is 1.69 square hours, and we have based our calculated average on thirty "samples." Thus, the expected squared deviation of these measured averages from the "ideal" average is 0.06 square hours.

At this point we are again faced with the unfortunate units of square hours. And, as with the variance, the solution is simply to take the square root, leading to our final answer:

$$\sigma_{\overline{X}} = \text{standard error} = \sqrt{E\left[(\overline{X} - \mu_x)^2\right]}$$
$$= \frac{\sigma_X}{\sqrt{N}}. \tag{4.24}$$

Numerically, the standard error is equal to the standard deviation of the probability distribution from which the samples are drawn, divided by the square root of the number of samples. For our example dealing with wrestling fringeheads,

$$\text{standard error of daily time wasted} = \sqrt{\frac{1.69}{30}} \cong 0.24. \tag{4.25}$$

In practical terms, the standard error tells us how much variation we can expect when we try to estimate experimentally the actual average of a random variable. The Central Limit Theorem tells us that our empirical averages are normally distributed. Thus, we can expect approximately 68% of our measured averages to lie within ±0.24 hours of the "true" mean.

Consider the following exercise. Let's assume that we can somehow replicate the experiment with our fringeheads so that there are one hundred individuals defending shelters on widely separated parts of the reef, and each individual independently calculates a daily average time that he wastes based on one month's worth of data. Due to the random nature of this experiment, these one hundred estimates of the mean are likely to differ one from the other, but the standard deviation of these means will be approximately equal to the standard error that we have calculated here, and sixty-eight of these one hundred estimates are likely to fall within 0.24 hours of the "true" daily average. If instead of one hundred individuals we had an infinite number, the standard deviation of the resulting means would exactly equal the standard error.

How could we get a better empirical estimate of the daily time wasted? Eq. (4.24) suggests two possibilities. First, reducing the variance in the daily time wasted would reduce the standard error. So, if we could control the day-to-day variation in the number of times a fish loses a wrestling match, we could at the end of the month obtain a more reliable estimate of the average daily time wasted. But then, if we could control how many times our fish win, we probably wouldn't have to deal with this problem at all. In many cases, reduction of the intrinsic variation in a process may not be practical. Alternatively, we can increase the number of samples that go into our calculation of the mean. For instance, if we keep track of the time wasted for two months instead of one, the standard error in our calculated mean is

$$\text{standard error (2 months)} = \sqrt{\frac{1.69}{60}} \cong 0.17,$$

71% of the value for thirty samples. Because the standard error varies with $1/\sqrt{N}$, doubling the number of samples reduces the likely error by about 29%.

As you move through this book, many cases will arise in which the reliability of an experimental estimate varies inversely with the square root of the number of samples taken. In each case, it may be helpful to refer back to the calculations we have just made to review the logic behind the standard error.

One final note: in deriving an expression for the standard error, we never needed to make reference to a particular probability density function. In other words, the definition of the standard error, $\sigma_X/\sqrt{N}$, applies to *any* probability distribution. All you need to be able to do is calculate σ_X from the underlying normal distribution. But what happens if σ_X is unknown? Read on.

4.7 *A Brief Detour to Statistics*

In the last four chapters we have explored the patterns of variability found in chance events with the goal of developing mathematical tools that allow us to gain insight into how probabilistic events work. Having reached that goal, we now stand at a fork in the road. If we choose to turn to the right, we will find ourselves using this knowledge in a predictive fashion to explore a wide range of biological phenomena. This is the path that we will follow. We could just as easily turn left, however, into the realm of statistical inference in which we would use our new knowledge to test hypotheses. All we would have to do to make that turn would be to shift our perspective a bit. As we warned you in the Preface, we have chosen not to head down the statistics path very far, but it will be useful to take a short excursion along that road. If you want to pursue the field of inferential statistics in depth, you should consult the many excellent texts on the subject. For example, this eclectic collection ought to include something for any biologist's taste: Gonick and Smith (1993), Neter et al. (1996), Norman and Steiner (1993), Sokal and Rohlf (1995), Williams (1993), and Zar (1996).

Many questions involve comparing the value of some parameter among groups or with a theoretical prediction. For example, if we measure the time wasted by several random fringeheads on a reef near Monterey, California, we can calculate an average for the reef. A colleague down the coast might do the same for a different group of fringeheads and calculate a slightly different average. We might hypothesize that these average times are different because the two groups of fish come from different genetic stocks or face a different size distribution of intruders. Alternatively, the fish on the two reefs could actually be wasting similar amounts of time wrestling, and the difference in measured averages might be due solely to chance. Since we are only taking a small sample of fish from each reef, variation among reefs could easily be due to inaccuracies in our ability to estimate the true reef average from a small sample. How can we decide which of these two alternatives is most probable? This is the job of inferential statistics.

One approach is to estimate the probability that two samples with different mean values could have been drawn by chance from the same underlying population. As we have seen in this chapter, the Central Limit Theorem states that the means of samples drawn from any probability distribution vary in a predictable way: the distribution of the sample means is described by the normal distribution. Furthermore, we have shown that the mean of the sample means is equal to the expectation, and the standard deviation of sample means is equal to the standard error, $\sigma_X/\sqrt{N}$. Thus, if we know the mean and standard deviation of the underlying probability distribution, we know what the distribution of sample means should be.

What if we turn this logic on its head? If we have a sample with a particular average, what can we say about the mean of the unknown distribution from which it was taken? We know that the two means (the sample mean and the population mean) are unlikely to be exactly the same under any circumstance (especially if the sample is small), but how different are they likely to be? Well, as we have just noted, the standard deviation of sample means (the standard error) measures the variability of sample means around the true population mean. If the standard error is very small, the distribution of sample means will be very narrow. As a result, the true mean is unlikely to be very far from any sample estimate. By contrast, if the standard error is large, the distribution of sample means will be broad, and the true mean will commonly differ greatly from the mean of a single sample. The probability distribution of sample means therefore gives us a way to estimate the position of the true population mean. In practice, this provides a way to estimate our confidence that a particular sample came from a particular population.

To see how this works, let's return to our discussion of the normal distribution, in which we noted that approximately 95% of values lie within two standard deviations of the mean. As a result, because we know that the standard error is the standard deviation of sample means, we also know that 95% of the sample means lie with two standard errors of the population mean. The flip side of this statement is that if we create an interval ±2 standard errors around each of a series of sample means, 95% of the time this region will include the unknown true population mean. This region around the sample mean is called a *confidence interval*, in this case a 95% confidence interval. The confidence interval for a particular parameter (such as the mean) is estimated from sample data in such a way that it defines an interval where there is a specified probability (e.g., 95%) that the true parameter value will occur within the interval. It would thus seem that the normal distribution provides an easy way to estimate our confidence in sample means.

There is just one problem, and unfortunately it is a big problem. Our estimate of the standard error assumes that we know σ_X, the standard deviation of values within the unknown probability distribution of the entire population. If we don't know the mean of the distribution, how can we know σ_X? Well, in virtually all circumstances we will *not* know σ_X. We can, however, estimate it.

When we take a sample (even a small one), there is likely to be variation among the individual values. The extent of this variation depends on the variation present in the population and thereby provides a source of information about σ_X. Fortunately, the *sample standard deviation*,

$$ s_X = \sqrt{\frac{\sum_{i=1}^{N}\left(x_i - \overline{X}\right)^2}{N-1}}, \tag{4.26} $$

is in many cases a reasonable estimate of σ_X itself. The standard error (the standard deviation of sample means) can then be estimated as

$$ \sigma_{\overline{X}} \cong s_{\overline{X}} = \frac{s_X}{\sqrt{N}}. \tag{4.27} $$

If N is large (> 20), the distribution of sample means converges on a normal distribution, and $s_{\overline{X}}$ can be substituted for $\sigma_{\overline{X}}$ when calculating confidence intervals. For smaller sample sizes, however, the distribution of sample means is no longer normal. In general, sample means have a *Student's t distribution*, a cousin of the normal distribution. The t distribution superficially looks like the normal distribution, but its shape changes slightly with N, and the changes are particularly noticeable when N is small. The details of this distribution are thoroughly addressed in nearly every statistics book.

This process we have just outlined (taking an estimate from a random sample and constructing a confidence interval to define a range of values where the true parameter value is likely to occur) is the basis of much of inferential statistics. Obviously, the details can get quite messy (otherwise statistics books would be very short), but the approach is relatively simple. Comparing a sample value to a theoretical prediction involves seeing if the predicted value lies within the confidence interval of the sample. Comparing two sample means to see if they indicate true underlying population differences involves looking at the pattern of overlap in the confidence intervals constructed around the two means. More complex questions (such as comparisons among a large number of samples) lead to much more complicated calculations, but at their core they follow a similar approach: (1) make inferences about population values on the basis of sample values; (2) examine hypotheses using these population estimates.

That is enough of a detour. Let's return to our mission. At long last we are in a position to apply the theory of probability in a *predictive* fashion to biology, and it is to this exploration that we turn in chapter 5.

4.8 *Summary*

By shifting our focus from individual outcomes to intervals of values, we have developed methods for quantifying the probabilities associated with *continuous distributions*. Our primary tools are the *cumulative probability distribution*, which allows us to calculate the probability of encountering a value less than or equal to some specified number, and the *probability density function*, which is analogous to the probability distributions we dealt with in chapter 3. The cumulative probability distribution is the integral of the probability density function. The *normal curve*, a particular example of a probability density function, is widely encountered in nature and will be of particular use in the rest of this book. We have explored its connection to the binomial distribution, and have tabulated its cumulative probability. And finally, we developed the concept of the *standard error*, an estimate of the variability among means. With these tools now in hand, we are ready to immerse ourselves in the study of chance in biology.

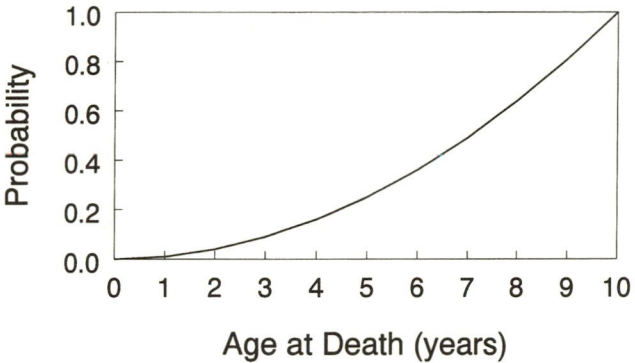

FIG. 4.13 The hypothetical cumulative probability function for a population of fish.

4.9 *Problems*

1. The state of Wyoming (U.S.A.) lies between longitudes 104°W and 111°W, and between latitudes 41°N and 45°N. Meteorites strike the Earth at random with respect to space and time. What is the probability that the next meteorite that arrives will strike Wyoming on a Tuesday if:

a. We know in advance that the meteorite is traveling along a path parallel to the Earth's axis and will arrive from the north?

b. The direction of the meteorite's approach is purely a matter of chance? In this case, there is equal probability that the meteorite will strike any location on Earth.

2. In a hypothetical population of fish, the cumulative probability of dying follows the curve shown in figure 4.13. All fish die by the time they are 10 years old, and the equation for the curve shown is

$$P(\text{age at death} \leq x) = \left(\frac{x}{10}\right)^2.$$

a. What is the probability density function for this population?

b. How likely is it that a fish will die between the ages of 3 and 4 years?

3. You are working with a population of sunflowers whose heights are normally distributed with a mean $\mu = 2.3$ meters and a standard deviation $\sigma = 0.5$ meters.

a. If you pick a plant at random from this population, what is the probability that the height of this sample is between 2.6 and 2.9 meters?

b. If you pick a plant at random, what is the probability that its height is *exactly* 1.9 meters?

4. In one thousand throws of a die, what is the approximate probability that a 3 is obtained fewer than 145 times?

5. For a population of sticklebacks, the length of the dorsal spine is normally distributed with a mean of 14 mm and a standard deviation of 3 mm. In a random sample of spines from ten fish drawn from this population, what is the probability that the mean of the *sample* is greater than 15 mm? Work the problem again for a sample size of a hundred (rather than ten) fish.

4.10 *Appendix 1: The Normal Distribution*

In eq. (4.17) we asserted that for large n, a binomial distribution can be approximated by the probability density function

$$g(X \mid \mu_X, \sigma_X) = \frac{1}{\sigma_X \sqrt{2\pi}} \exp\left[-\frac{1}{2} \left(\frac{x - \mu_X}{\sigma_X} \right)^2 \right]. \tag{4.28}$$

But this formula bears no resemblance to that of a binomial distribution:

$$P(S = i) = \frac{n!}{i!(n - i)!} \cdot p^i q^{n-i}. \tag{4.29}$$

How did Bernoulli and De Moivre work from one equation to the other? Surprisingly, very few statistics books provide the answer; the reader is generally asked to take it on faith that the equation for a normal probability density function behaves as advertised. Being the skeptical sort, however, we feel compelled to back up our claim, and we provide below an informal derivation of the equation for the normal distribution. The approach here is that of Ruhla (1992).

We begin, as usual, by defining our terms. We examine a binomial distribution based on a series of Bernoulli trials. Each trial has a probability of success p and a probability of failure $q = 1 - p$. For simplicity, we assume that the value of success is 1 and the value of failure is 0. We are interested in the shape of the binomial distribution when the number of individual trials, n, is very large. Furthermore, we specify that i, the number of successes, is also large, but small compared to n. In other words, $1 \ll i \ll n$. We now make a calculation out of the blue that will prove useful later on. If i (the number of successes we encounter in n Bernoulli trials) is very large, a change of i to $i + 1$ or $i - 1$ is very small compared to n, and we can reasonably treat i as if it were a continuous variable. This allows us to calculate the following derivative:

$$\frac{d(\ln i!)}{di} \cong \frac{\ln(i + 1)! - \ln i!}{1}, \tag{4.30}$$

in which we have taken the infinitesimal di to be equal to the smallest value available, 1. This expression can be simplified:

$$\frac{\ln(i+1)! - \ln i!}{1} = \ln \frac{(i+1)!}{i!} = \ln(i+1). \tag{4.31}$$

But then, because i is so large, $i + 1 \cong i$, leading to the conclusion that for large i,

$$\frac{d(\ln i!)}{di} = \ln i. \tag{4.32}$$

We will make use of this relationship presently. Now, on with the derivation itself.

We start by taking the logarithm of the binomial distribution:

$$\ln P(i) = \ln n! - \ln i! - \ln(n-i)! + i \ln p + (n-i) \ln q. \tag{4.33}$$

For what value of i is $\ln P(i)$ maximal? (It is probably not at all clear why one would ask this question, but bear with us; it leads to interesting results.) To search for the maximum of $\ln P(i)$, we take its derivative and set the derivative equal to 0. That is,

$$\frac{d \ln P(i)}{di} = 0. \tag{4.34}$$

Carrying out the differentiation,

$$\frac{d \ln P(i)}{di} = 0 - \frac{d \ln i!}{di} - \frac{d \ln(n-i)!}{di} + \ln p - \ln q. \tag{4.35}$$

We now make use of the approximation established in eq. (4.32) to simplify this expression:

$$\frac{d \ln P(i)}{di} = -\ln i + \ln(n-i) + \ln p - \ln q \tag{4.36}$$

$$= \ln\left(\frac{n-i}{i}\frac{p}{q}\right). \tag{4.37}$$

Setting this equal to 0 (to solve for the maximum), we conclude that

$$\frac{n-i}{i}\frac{p}{q} = 1. \tag{4.38}$$

A bit of algebra (and the knowledge that $q = 1 - p$) leads to the conclusion that $\ln P(i)$ (and therefore $P(i)$ itself) is maximal when $i = np$. Recall, however, that np is the expected (or mean) value of the binomial distribution, which we can call μ_X. Thus, so far, we have established that the most probable i and the expected i are one and the same for the binomial distribution when n is large.

But what shape is the distribution itself? As a first approximation, we express the logarithm of the distribution as a series expansion around its mean value μ_X:

$$\ln P(i) \cong \ln P(\mu_X) + \frac{d \ln P(\mu_X)}{di}(i - \mu_X)$$

$$+ \frac{1}{2}\frac{d^2 \ln P(\mu_X)}{di^2}(i - \mu_X)^2 \ldots \quad (4.39)$$

We could continue the expansion to higher-order terms, but these few will suffice. We can immediately simplify this equation by noting (from our calculations above) that $d \ln P(\mu_X)/di = 0$, so the second term on the right-hand side of the equation vanishes. We are left, then, to calculate the second derivative in the third term. Recalling that

$$\frac{d \ln P(i)}{di} = -\ln i + \ln(n - i) + \ln p - \ln q, \quad (4.40)$$

and crunching through the second differentiation, we see that

$$\frac{d^2 \ln P(i)}{di^2} = -\frac{1}{i} - \frac{1}{n - i} = \frac{n}{i(n - i)}. \quad (4.41)$$

Setting $i = \mu_X$ and again recalling that $\mu_X = np$ (chapter 3):

$$\frac{d^2 \ln P(\mu_X)}{di^2} = \frac{n}{np(n - np)} = \frac{1}{np(1 - p)}. \quad (4.42)$$

Now $1 - p = q$, so

$$\frac{d^2 \ln P(\mu_X)}{di^2} = \frac{1}{npq}. \quad (4.43)$$

But we can take this one step farther. Recall that npq is the variance of the binomial distribution, σ_X^2 (eq. 3.59). Thus,

$$\frac{d^2 \ln P(\mu_X)}{di^2} = \frac{1}{\sigma_X^2}. \quad (4.44)$$

Inserting this result back into eq. (4.39), we arrive at an equation describing the shape of the logarithm of the probability distribution:

$$\ln P(i) = \ln P(\mu_X) - \frac{(i - \mu_X)^2}{2\sigma_X^2}. \quad (4.45)$$

Taking the antilog of each side of this equation, we see that

$$P(i) = P(\mu_X) \exp\left[-\frac{1}{2}\frac{(i-\mu_X)^2}{\sigma_X^2}\right]. \tag{4.46}$$

This looks promising. All we have to do now is calculate the probability of μ_X. To do so, we return to one of the basic rules of probability:

$$\sum_{i=0}^{n} P(i) = 1. \tag{4.47}$$

But because we are treating i as a continuous variable, we can replace this summation with the analogous integral:

$$\int_0^n P(i)di = 1. \tag{4.48}$$

Now, $P(i)$ will have appreciable values only in a narrow range near μ_X, the position around which we have taken our series expansion. As a result, we are justified in extending the range of our integration to infinity, with the implicit understanding that $P(i) = 0$ when $i < 0$ or $i > n$. Thus,

$$\int_{-\infty}^{\infty} P(i)di = 1. \tag{4.49}$$

Inserting our calculated expression for $P(i)$ then gives us

$$\int_{-\infty}^{\infty} P(\mu_X) \exp\left[-\frac{1}{2}\frac{(i-\mu_X)^2}{\sigma_X^2}\right]di = 1. \tag{4.50}$$

At this point it will help to streamline our equation by substituting $u = (i - \mu_X)/\sqrt{2}\sigma_X$. With this substitution in place,

$$\int_{-\infty}^{\infty} P(\mu_X)(\sqrt{2}\sigma_X) \exp(-u^2)du = 1 \tag{4.51}$$

$$P(\mu_X)(\sqrt{2}\sigma_X) \int_{-\infty}^{\infty} \exp(-u^2)du = 1. \tag{4.52}$$

Now, $\int_{-\infty}^{\infty} \exp(-u^2)du = \sqrt{\pi}$ (trust us, or you can look it up!). Thus,

$$P(\mu_X) = \frac{1}{\sigma_X\sqrt{2\pi}}, \tag{4.53}$$

and

$$P(i) = \frac{1}{\sigma_X\sqrt{2\pi}} \exp\left[-\frac{1}{2}\frac{(i-\mu_X)^2}{\sigma_X^2}\right]. \tag{4.54}$$

We are very close to our final answer, but one step remains. So far we are still dealing with a probability distribution, rather than a probability density function. We need to make the transition from our integer variable i to a continuous variable, x. To do this we choose an interval in x, $[x, x + \Delta x]$ such that $1 \ll \Delta x \ll i$. We can do this because (you will remember) we have assumed that i is very large. Now, given this constraint on Δx, there must be Δx values of i in our interval (give or take 1). If there are Δx values of i in the interval, the overall probability of finding i in this interval is (to a close approximation) $P(i)\Delta x$. Or, in terms of our continuous variable x, we can write this probability as

$$g(X)\Delta x = \frac{1}{\sigma_X \sqrt{2\pi}} \exp\left[-\frac{1}{2} \frac{(x - \mu_X)^2}{\sigma^{X^2}}\right] \Delta x. \qquad (4.55)$$

And finally,

$$g(X) = \frac{1}{\sigma_X \sqrt{2\pi}} \exp\left[-\frac{1}{2} \frac{(x - \mu_X)^2}{\sigma_X^2}\right]. \qquad (4.56)$$

This is the normal probability density function. Rewriting this equation with the transformed variable $Z = (x - \mu_X)/\sigma_X$ results in the *standard* normal probability density function:

$$g(Z) = \frac{1}{\sqrt{2\pi}} \exp\left(-\frac{z^2}{2}\right). \qquad (4.57)$$

As De Moivre might have remarked, voilá!

4.11 *Appendix 2: The Central Limit Theorem*

The Central Limit Theorem is a keystone of statistics. It asserts that a random variable that is itself the sum of a large number of other random variables will be normally distributed regardless of the distribution of the variables from which it is constructed. This property goes far to explain why the normal distribution appears so often in nature, but it almost seems too good to be true. How can the distribution of the sum be independent of the distribution of its randomly varying parts? What follows here is an informal demonstration of the validity of the Central Limit Theorem, following the logic of Reif (1965).

Consider a nest of seed-gathering ants. During the summer, worker ants forage for grass seeds, attempting to build up sufficient stock to last the colony through the winter. In a year, the workers gather N seeds. Through diligent observation of the nest we are able to record the weight of each individual seed. The weight of the first seed brought into the nest is s_1, the weight of the second seed is

s_2, etc. The weight of the Nth seed is s_N. At the end of the summer, the total weight of seeds accumulated is

$$x = \sum_{i=1}^{N} s_i. \qquad (4.58)$$

Having weighed each of the N seeds, we can accurately quantify the accumulated weight for this particular nest in this particular summer. As scientists, however, we wish to make the most of our data and desire to use our observations to describe in general the weight of seeds collected by a colony of ants. But in attempting to make this generalization we run into a familiar problem: there is some variation among the weights of individual seeds, and the seeds brought into the nest are chosen at random. As a result, our observation of the weight accumulated by our particular nest is merely one realization of a random process. In other words, the value x we obtain in this particular case is one example of the random variable X, the weight of seeds accumulated. If we were to repeat our observations on other nests, we would in all likelihood obtain a different value of X. What is the probability that, for a nest chosen at random, X lies between x and $x + dx$? To answer this question, we need to know $g(X)$, the probability density function of X. The calculation of this function is our task here.

We begin by establishing a few conditions. The variation in weight of individual seeds is itself described by a probability density function, $w(s)$; $w(s)$ is the same for all seeds. By definition, then, the probability that the weight of the first seed brought into the nest lies in the small range ds_1 is $w(s_1)ds_1$. The probability that the weight of seed 2 lies in the range ds_2 is $w(s_2)ds_2$, and so forth. We assume that the weight of any individual seed is independent of the weight of other seeds.

Given these conditions, we can then assert that for any particular sequence of seeds brought into the nest such that

the weight of seed 1 lies between s_1 and $s_1 + ds_1$, and
the weight of seed 2 lies between s_2 and $s_2 + ds_2$, and ...
the weight of seed N lies between s_N and $s_N + ds_N$,

the overall probability that this particular sequence should occur is the product of the individual probabilities associated with each seed (yet another application of the product rule, chapter 2):

probability of this sequence $= w(s_1)ds_1 \times w(s_2)ds_2 \times \ldots w(s_N)ds_N.$ (4.59)

As we know, this particular sequence of seeds results in $x < X < x + dx$. But a variety of other sequences may have the same result. Only if we can sum the probability across all sequences that yield $x < X < x + dx$ will we have our answer. In other words, we desire to solve the integral

$$\text{probability} = g(X)dx$$
$$= \int \int \ldots \int w(s_1)w(s_2)\ldots w(s_N)ds_1 ds_2 \ldots ds_N, \quad (4.60)$$

where it is understood that the integration is carried out over all conditions in which

$$x < \sum_{i=1}^{N} s_i < x + dx. \quad (4.61)$$

To perform this multiple integration appropriately, one would have to specify carefully the limits for each individual integration such that the condition of eq. (4.61) is met. Exactly how one would juggle the limits to accomplish this is not at all clear.

Fortunately, there is a trick that avoids this problem. In essence, we specify the condition given by eq. (4.61) by fiddling with the integrand (the expression being integrated) rather than with the limits to integration. If we do this right, it will allow us to evaluate each integral over the general range $-\infty$ to ∞ and still get the correct answer. The necessary adjustment to eq. (4.60) is made using the Dirac δ function, $\delta(x - x_0)$. This handy function has the properties

$$\delta(x - x_0)dx = 0 \quad \text{if } |x - x_0| > \frac{dx}{2},$$
$$\delta(x - x_0)dx = 1 \quad \text{if } |x - x_0| < \frac{dx}{2}, \quad (4.62)$$

which allow us to restate eq. (4.60) as

$$g(X)dx = \int_{-\infty}^{\infty} \int_{-\infty}^{\infty} \ldots \int_{-\infty}^{\infty} w(s_1)w(s_2)\ldots w(s_N)$$
$$\times \left[\delta\left(x - \sum_{i=1}^{N} s_i \right) dx \right] ds_1 ds_2 \ldots ds_N. \quad (4.63)$$

The presence of the δ function elegantly ensures that each integral only contributes to $g(X)$ when the condition of eq. (4.61) is met.

This is all very neat, but we are still left with an extremely unwieldy multiple integral. To proceed, we take advantage of an additional property of the Dirac δ function. It can be shown (see Reif 1965, for example) that

$$\delta(x - x_0) = \frac{1}{2\pi} \int_{-\infty}^{\infty} \exp\left[ik(x_0 - x) \right] dk. \quad (4.64)$$

Here i is the square root of -1, and k is a dummy variable used solely to carry out the integration. Substituting this equality into eq. (4.63) and dividing through by dx, we see that

$$g(X) = \int_{-\infty}^{\infty} \int_{-\infty}^{\infty} \cdots \int_{-\infty}^{\infty} w(s_1)w(s_2)\ldots$$

$$\times\, w(s_N)\frac{1}{2\pi}\int_{-\infty}^{\infty} \exp[ik(s_1 + s_2 \ldots s_N - x)]dk\,ds_1 ds_2 \ldots ds_N. \quad (4.65)$$

Making use of the fact that $\exp[a + b] = \exp[a] \cdot \exp[b]$, we can rearrange this expression:

$$g(X) = \frac{1}{2\pi}\int_{-\infty}^{\infty} \exp[-ikx]dk \int_{-\infty}^{\infty} w(s_1)\exp[iks_i]ds_1 \ldots$$

$$\times \int_{-\infty}^{\infty} w(s_N)\exp[iks_N]ds_N. \quad (4.66)$$

Recall that $w(s_1) = w(s_2)\ldots = w(s_N)$. As a consequence, the last N integrals in this expression are functionally the same, and we can streamline our notation by giving them their own symbol:

$$Q(k) \equiv \int_{-\infty}^{\infty} w(s)\exp[iks]ds. \quad (4.67)$$

Thus,

$$\int_{-\infty}^{\infty} w(s_1)\exp[iks_i]ds_1 \ldots \int_{-\infty}^{\infty} w(s_N)\exp[iks_N]ds_N$$

$$= Q_1(k) \cdot Q_2(k)\ldots Q_N(k)$$

$$= Q^N(k), \quad (4.68)$$

and

$$g(X) = \frac{1}{2\pi}\int_{-\infty}^{\infty} \exp[-ikx]Q^N(k)dk. \quad (4.69)$$

If we can evaluate this expression, we have our answer.

Solving eq. (4.69) exactly would be problematic, but we can arrive at a close approximation if we assume that N is large. The argument proceeds as follows.

We first examine the properties of the expression $\exp[iks]$ found in $Q(k)$. A quick glance at any standard calculus text will refresh your memory that

$$\exp[iks] \equiv \cos(ks) + i\sin(ks). \quad (4.70)$$

In other words, both the real and the imaginary parts of this expression oscillate around zero, making one complete cycle as ks goes from 0 to 2π. If we were to record the values of exp[iks] over a small fraction of a cycle, it is quite possible that the average of these values would be different from zero. However, if we continue to record values for one complete cycle (or for many cycles), the average of this function is approximately equal to zero. Now, the rate of oscillation of exp[iks] depends on the magnitude of k. When k is large, exp[iks] varies rapidly over a small change in s (over the range from a to b, for instance). This allows us to make the following approximation. Over a range of s in which $w(s)$ changes relatively slowly ($|dw(s)/ds|/k \ll w(s)$),

$$\int_a^b w(s) \exp[iks]ds \cong w(s) \int_a^b \exp[iks]ds \cong 0. \tag{4.71}$$

This implies that for large values of k, $Q(k) \cong 0$, and when N is large, $Q^N(k)$ is very much smaller still. As a result, when we strive to evaluate the integral in eq. (4.69), we need do so only for small values of k; as long as N is large the integrand is effectively zero when k is large.

The value of this conclusion is that it allows us to approximate $Q(k)$ for small k using a Taylor series expansion of k:

$$Q(k) \equiv \int_{-\infty}^{\infty} w(s) \exp[iks]ds$$
$$= \int_{-\infty}^{\infty} w(s)(1 + iks - \frac{1}{2}k^2s^2 + \ldots)ds. \tag{4.72}$$

This in turn can be rewritten as

$$Q(k) = \int_{-\infty}^{\infty} w(s)ds + ik \int_{-\infty}^{\infty} sw(s)ds - \frac{1}{2}k^2 \int_{-\infty}^{\infty} s^2w(s)ds + \ldots. \tag{4.73}$$

We assume that $w(s)$ rapidly approaches zero as $|s| \longrightarrow \infty$, in which case each of the integrals in this expression remains finite. Note that this assumption is hardly restrictive. Although there is some variation in the weight of grass seeds, we can reasonably expect that the probability of encountering a seed of a particular weight decreases rapidly as weight deviates much from the mean.

The integrals in eq. (4.73) should look familiar. We know from the properties of any probability distribution that $\int_{-\infty}^{\infty} w(s)ds = 1$. From the definition of the mean we know that $\int_{-\infty}^{\infty} sw(s)ds = \bar{s}$. By the same token, $\int_{-\infty}^{\infty} s^2w(s)ds = \overline{s^2}$. Thus,

$$Q(k) = 1 + ik\bar{s} - \frac{1}{2}k^2\overline{s^2}\ldots. \tag{4.74}$$

As a practical matter, because $Q(k)$ oscillates so fast as a function of k, it is advantageous to work instead with the logarithm of $Q(k)$, which varies more slowly:

$$\ln Q^N(k) = N \ln Q(k) = N \ln\left(1 + ik\overline{s} - \frac{1}{2}k^2\overline{s^2}\dots\right). \tag{4.75}$$

We now expand *this* term, again using a Taylor series. Noting that for $y \ll 1$,

$$\ln(1 + y) = y - \frac{1}{2}y^2\dots, \tag{4.76}$$

we can write eq. (4.75) as

$$\ln Q^N = N\left[ik\overline{s} - \frac{1}{2}k^2\overline{s^2} - \frac{1}{2}(ik\overline{s})^2\dots\right], \tag{4.77}$$

where we have retained only terms up to those containing k^2. Because we assume that k is small, terms containing higher powers of k are *very* small and we can safely ignore them. Rearranging terms, we see that

$$\ln Q^N = N\left[ik\overline{s} - \frac{1}{2}k^2\left(\overline{s^2} - \overline{s}^2\right)\dots\right]. \tag{4.78}$$

The expression $\overline{s^2} - \overline{s}^2$ may look familiar. Recall from chapter 3 (Corollary 3.1) that the difference between the mean square of a variable (in this case $\overline{s^2}$) and the mean of the variable squared ($\overline{s}^2$) is the variance. Thus, if we define Δs as the deviation of s from its mean, the variance of s, $\overline{(\Delta s)^2} \equiv \overline{s^2} - \overline{s}^2$ and

$$\ln Q^N = N\left[ik\overline{s} - \frac{1}{2}k^2\overline{(\Delta s)^2}\dots\right]. \tag{4.79}$$

Taking the antilog of this expression we arrive at the conclusion that

$$Q^N(k) = \exp\left[iNk\overline{s} - \frac{1}{2}Nk^2\overline{(\Delta s)^2}\right], \tag{4.80}$$

and from eq. (4.69),

$$g(X) = \frac{1}{2\pi}\int_{-\infty}^{\infty} \exp\left[ik(N\overline{s} - x) - \frac{1}{2}Nk^2\overline{(\Delta s)^2}\right]dk. \tag{4.81}$$

If we could just evaluate this integral we would be done. To this end we consult a table of integrals and note that eq. (4.81) has a standard form. Given that b is a positive real number,

$$\int_{-\infty}^{\infty} \exp(ak - bk^2)dk = \int_{-\infty}^{\infty} \exp\left[-b\left(-\frac{a}{b}k + k^2\right)\right]dk$$

$$= \int_{-\infty}^{\infty} \exp\left\{-b\left[\left(-\frac{a}{2b} + k\right)^2 - \frac{a^2}{4b^2}\right]\right\}dk$$

$$= \int_{-\infty}^{\infty} \exp\left\{-b\left(-\frac{a}{2b} + k\right)^2 + \frac{a^2}{4b}\right\}. \tag{4.82}$$

By substituting $y = -a/2b + k$, this becomes

$$\int_{-\infty}^{\infty} \exp(ak - bk^2)dk = \exp\left(\frac{a^2}{4b}\right) \int_{-\infty}^{\infty} \exp(-by^2)dy. \tag{4.83}$$

It so happens that $\int_{-\infty}^{\infty} \exp(-by^2)dy = \sqrt{\pi/b}$ (again from a consultation with the integral tables). Thus,

$$\int_{-\infty}^{\infty} \exp(ak - bk^2)dk = \sqrt{\frac{\pi}{b}} \exp\left(\frac{a^2}{4b}\right). \tag{4.84}$$

This is a formula we can now apply to the task at hand. Setting $a = i(N\overline{s} - x)$ and $b = 1/2N\overline{(\Delta s)^2}$ and recalling that $i^2 = -1$,

$$g(X) = \sqrt{\frac{\pi}{\frac{1}{2}N\overline{(\Delta s)^2}}} \exp\left[-\frac{(N\overline{s} - x)^2}{4 \cdot \frac{1}{2}N\overline{(\Delta s)^2}}\right]. \tag{4.85}$$

One more set of substitutions,

$$\mu_X \equiv N\overline{s}$$

$$\sigma_X^2 \equiv N\overline{(\Delta s)^2}, \tag{4.86}$$

and all of the sudden we arrive at our destination:

$$g(X) = \frac{1}{\sigma_X\sqrt{2\pi}} \exp\left[-\frac{(x - \mu)^2}{2\sigma_X^2}\right]. \tag{4.87}$$

In other words, the probability density function of X (X = the summed weight of seeds) has the form of a *normal distribution* with a mean equal to N times the mean of individual seeds and a variance equal to N times the variance of individual seeds. Recall that we began this exercise by assuming that the weight of seeds had a probability density function of its own, $w(s)$, but at no point did we specify the shape of this function. Thus, the conclusion that we have just reached (the distribution of the sum is normal) must hold no matter what exact shape $w(s)$ may have. All we have required is that the weight of each individual seed is independent of the weights of others and that $w(s)$ rapidly approaches zero as $|s| \longrightarrow \infty$.

We have couched this example in terms of ants gathering seeds, but it can easily be translated to any of an exceptionally wide variety of situations. Thus, in general, any random variable that is itself the sum of other independent random variables can be described by a normal probability density function. Therein lies the essence of the Central Limit Theorem.

5

Random Walks

In this chapter and the next we make our first practical, scientific use of the probability distributions introduced in chapters 3 and 4. We will show you how the binomial and normal distributions can be used to explain phenomena as far-ranging as molecular diffusion and the rates at which phytoplankton are mixed in tropical lagoons, how the genetic composition of a population can drift across generations, and why your arteries are rubbery. All of these subjects depend on the statistics of *random walks*. We begin with a discussion of molecular diffusion.

5.1 *The Motion of Molecules*

It is commonly understood that heat is a form of energy. For instance, heat given off by a fire can convert water to steam, and the expanding vapor can be used to propel a steam engine, turn a turbine, or perform other types of work. In an analogy from nature, heat from the sun evaporates water and powers the wind, and these combined processes can lead to energetic phenomena such as hurricanes and tornadoes. What is less commonly understood is the subtle way in which heat is different from all other types of energy. The difference is best grasped through an example.

Consider the motion of a single molecule. If the molecule is part of a solid (an iron molecule in a bolt, or a calcium molecule in your teeth), its motion is constrained. The molecule can rattle about a bit, but its bonds to adjacent molecules ensure that it is not free to move very far. These constraints are relaxed for a molecule in a fluid, for instance a molecule of oxygen in the air that you are breathing. If the molecule has mass m and moves at a speed u, it has a kinetic energy

$$\text{kinetic energy} = \frac{mu^2}{2}. \tag{5.1}$$

This energy, measured in joules, remains constant as the molecule travels until, by chance, the molecule runs into another molecule of gas. In the collision, each molecule is likely to change both its speed and direction, and while the total kinetic energy of the *pair* remains constant, the energy of each *individual* molecule can change.

Now, collisions among molecules are far from rare. In air at room temperature and sea level, each molecule of oxygen collides with a neighbor about 6 billion times per second. In water, where molecules are packed much closer together, collisions occur about 60 billion times per second (Denny 1993). Given this wild billiard game of molecular collisions, it would be very difficult to specify precisely what the kinetic energy of a given molecule is at any particular time. Faced with this problem, we take a practical approach—we concentrate on the *mean* kinetic energy of molecules rather than their instantaneous energy.

On average, how much kinetic energy does a molecule have? The answer turns out to depend on the temperature. The higher the temperature, the higher the average kinetic energy of a molecule. In fact, the relationship between temperature and average kinetic energy provides one of the standard definitions of temperature. The absolute temperature of a molecule, T, is defined as

$$T \equiv \frac{\overline{mu^2}}{3k}. \tag{5.2}$$

Here T is measured in kelvins, and k is Boltzmann's constant, 1.38×10^{-23} joules per kelvin (Reif 1965). The bar over mu^2 signifies that the kinetic energy is an average taken over a time sufficient to allow for many collisions.

This definition can be rearranged to make explicit the relationship between temperature and average kinetic energy:

$$\text{thermal kinetic energy} = \frac{\overline{mu^2}}{2} = \frac{3kT}{2}. \tag{5.3}$$

So, if we know the temperature, we can compute the thermal kinetic energy of a molecule. How is this any different from the kinetic energy of a thrown baseball or a moving truck? On the small spatial scale of a molecule in motion between collisions, there really is no difference. A molecule of a certain mass moves in a certain direction at a given speed, and therefore has a defined kinetic energy. However, the frequent changes in direction and speed that the molecule undergoes give its kinetic energy, when averaged over time and space, a set of properties that are fundamentally different from those we associate with, for instance, a baseball in motion.

The motion of the ball is *ordered* in the sense that all its molecules move in virtual synchrony, and it is possible to predict with some accuracy where the ball will be at some later time. This predictability is not true of an individual molecule of a gas or liquid. Because a molecule in a fluid collides so often, the direction and speed of its motion are constantly changing. And this random behavior makes it impossible to predict precisely where the molecule will be even a fraction of a second in the future.

The random motion of molecules has profound implications for physics and engineering. For example, while it is easy to accomplish the complete conversion of other forms of energy into heat, it is a practical impossibility to convert heat completely back into the form of energy whence it came. This one-way nature of heat governs much about how we live. When you exercise, your muscles convert to heat much of the chemical potential energy you ingested as food. The process does not work well in reverse, however. You cannot bask in the sun and expect the heat you absorb to be efficiently stored as chemical energy.

Other examples abound. In fact, the physics of heat is so important that a whole branch of science—thermodynamics—has been formed around its study, and the one-way nature of thermal energy is formally expressed as thermodynamics' second law (Atkins 1984). We will not delve deeply into the mysteries of thermodynamics. Instead, we will focus on the property of molecules that defines heat, that is, their random motion.

Equation (5.3) defines a relationship among temperature, molecular mass, and speed. We can use this relationship to calculate the average velocity of a molecule in a particular direction. Suppose we define a set of Cartesian coordinates in space; that is, we arbitrarily say that some point is the origin, and the x-, y-, and z-axes extend off in mutually perpendicular directions. Now, the motion of molecules is random, so whatever the directions we arbitrarily pick for our axes and the precise location of the origin, we can safely suppose that, on average, one-third of the kinetic energy of a molecule will be associated with each of the axes.[1] Given eq. (5.3) for the total kinetic energy, we divide by 3 to see that the average kinetic energy along the x-axis is:

$$\text{average kinetic energy along one axis} = \frac{kT}{2}. \qquad (5.4)$$

This expression allows us to calculate the average speed of a molecule along the axis. From the definition of kinetic energy (eq. 5.1), we know that the average kinetic energy of motion along the x-axis is

$$\text{average kinetic energy along the } x\text{-axis} = \frac{\overline{mu_x^2}}{2}, \qquad (5.5)$$

where u_x is the x-directed velocity. We have no reason to believe that the mass of the molecule changes through time, so we can remove the mass from the

[1] This property of random molecular motion is known as the *equipartition theorem*. For a thorough explanation, consult Reif (1965).

averaging process and rewrite the kinetic energy as

$$\text{average kinetic energy along the } x\text{-axis} = m\frac{\overline{u_x^2}}{2}. \tag{5.6}$$

Equating this expression with eq. (5.4), we find that

$$\overline{u_x^2} = \frac{kT}{m}. \tag{5.7}$$

The mean square velocity is equal to the ratio of kT to mass. Or, taking the square root of the mean square, we find that the average speed along the x-axis (the root mean square, or rms, speed) is

$$u_{x,\,\text{rms}} = \sqrt{\overline{u_x^2}} = \sqrt{\frac{kT}{m}}. \tag{5.8}$$

Consider the case of a molecule of everyday table sugar. The molecular weight of sucrose is 342 daltons. In other words, one mole of sucrose $(6.02 \times 10^{23}$ molecules) has a mass of 342 grams. In standard Système Internationale units, then, one mole of sucrose has a mass of 0.342 kilograms, and one molecule of sucrose has a mass of 5.7×10^{-25} kg. Inserting this value for m in eq. (5.8), we find that at the melting temperature of ice (273 K), the root mean square speed of a sucrose molecule along the x-axis is 81 ms^{-1}! Note that this conclusion is true no matter which axis we arbitrarily choose as our x-axis.

Eighty-one meters per second is a sizable speed. For example, if you pour a teaspoon of sugar into a glass of iced tea, it seems that it would hardly be worth stirring. After all, a molecule of sugar as it enters solution at the bottom of the glass could travel toward the top of the glass at a speed of 81 m s^{-1}. For a typical tea glass with a height of perhaps 0.15 m, the trip would take less than 0.002 seconds.

Why, then, in reality does it take substantial swizzling with a spoon to effectively mix sugar into a drink? The answer, as you probably suspect, has to do with the disordered motion of sugar molecules. Yes, at the temperature of your iced tea a molecule of sucrose can move at an average speed of 81 m s^{-1} toward the top of the glass, but it does so only until it hits a molecule of water, a frequent occurrence. The effect of these collisions is to randomize the direction in which the molecule moves, so that on average half the time the molecule is moving at great speed toward the ice cubes, and half the time it is moving with equal celerity toward the tablecloth. As a result, the *net* motion of the molecule as it follows this random walk is quite a bit slower than you might expect. Exactly how slow is something that we are now in a position to calculate.

5.2 *Rules of a Random Walk*

A sucrose molecule in a glass of tea, or an oxygen molecule in the air, is free to move in three dimensions. It is possible to calculate the statistics of a three-dimensional random walk; indeed, we will do so in the next chapter. But for simplicity at the start, we will initially examine the case of a random walk in a single dimension. In other words, we wish to follow the motion of a particle through time as it takes a random series of steps along the x-axis.

To begin the calculation, we need to define carefully the rules of the game. There are six of them:

1. We start at time $t = 0$ with the particle at $x = 0$.

2. The particle moves a fixed distance along the axis every τ seconds.

3. While the particle is moving, it moves at a velocity $\pm u_x$, so the effective step length, δ, is τu_x. Note that because δ depends on u_x, it in turn depends on temperature and the mass of the particle, the reason being that temperature and mass determine u_x (eq. 5.8).

4. The probability p of taking a step in the positive x direction (say, to the right) is $1/2$. By necessity, then, the probability q of taking a step to the left (in the negative x direction) is also $1/2$.

5. The direction of each step is independent of the previous step. That is, the particle has no "memory" of its past history and, in essence, flips a coin before each step to tell it in which direction to move.

6. We are free to place as many identical particles as we wish on the axis and they will nonetheless move independently of each other.

As we will see, these rules lead to an important, but nonintuitive, conclusion: the average distance traveled increases linearly with the *square root* of time. But to understand where this conclusion comes from, we must do a bit of mathematical preparation. As a start, let us first calculate what may appear to be a humdrum result, the average position of a particle on the axis after a given number of steps.

5.2.1 THE AVERAGE

There are two ways in which this calculation can proceed. We could place a single particle at the origin, turn it loose to follow a random walk, and after the prescribed number of steps note its position. We could then return the same particle to the origin and repeat the experiment N times. By summing the individual displacements and dividing by N, we could calculate the average displacement.

Alternatively, we could place N particles at the origin, release them all at the same time, and note their positions after the prescribed number of steps.

Summing the displacements of all particles and dividing by N again yields an average displacement. In this case, the average is termed an *ensemble average* because it is calculated for the group rather than the individual. In practice, because our rules specify that particles are identical and do not interact as they move, the ensemble and individual averages will be the same. We arbitrarily choose the ensemble method for our calculations.

But first, a bit of nomenclature. After n steps, each particle will be at a particular location on the x-axis. To keep track of where each particle is, we adopt the following notation: $x_i(n)$ is the position on the x-axis of the ith particle after n steps.

According to our rules, we know that the position of a particle after n steps differs from its position after $n - 1$ steps by the distance δ. Expressing this as an equation, we see that

$$x_i(n) = x_i(n - 1) + \delta, \tag{5.9}$$

where the expectation is that half the time δ is positive and half the time it is negative. Given this relationship, we can write an equation for the average displacement of N particles after n steps in terms of their average displacement after $n - 1$ steps:

$$\overline{X}(n) = \frac{1}{N} \sum_{i=1}^{N} x_i(n),$$

$$= \frac{1}{N} \sum_{i=1}^{N} [x_i(n - 1) + \delta],$$

$$= \frac{1}{N} \sum_{i=1}^{N} [x_i(n - 1)] + \frac{1}{N} \sum_{i=1}^{N} \delta. \tag{5.10}$$

Now, the last term in this expression, $\frac{1}{N} \sum_{i=1}^{N} \delta$, is the average distance moved by a single step, and for large N it asymptotes to the expectation of the step distance. If we treat each step as a Bernoulli trial with $p = q = 1/2$ and values δ (for a success) and $-\delta$ (for a failure), the expectation is

$$E(\text{step distance}) = p \cdot \delta + q \cdot -\delta$$

$$= \frac{1}{2} \delta - \frac{1}{2} \delta$$

$$= 0. \tag{5.11}$$

Thus,

$$\overline{X}(n) = \frac{1}{N} \sum_{i=1}^{N} x_i(n - 1) = \overline{X}(n - 1). \tag{5.12}$$

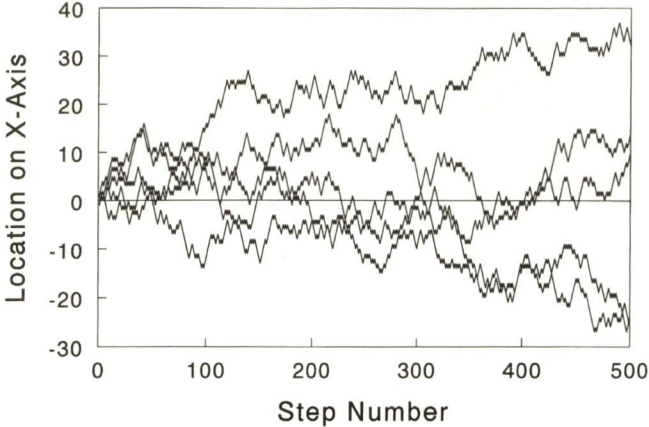

FIG. 5.1 Five particles, each released at the origin, spread out as they randomly walk along the x-axis. Although each particle is likely to wander from the origin, the average location of the ensemble of particles remains approximately at the starting point.

That is, the mean position after n steps is expected to be the same as the mean position after $n - 1$ steps. On average, the particle goes nowhere.

This should not be too surprising. Indeed, it is probably the conclusion you would have reached intuitively if you were told that the particle had an equal chance of moving left and right along the axis. There is comfort, however, in having the math confirm our intuition. It gives us some faith when, as we are about to see, the math makes predictions about the motion of particles that are nonintuitive.

5.2.2 THE VARIANCE

Consider a question closely related to (but subtly different from) the one we have just examined: Where, exactly, would we expect to find a given particle after a certain number of steps? We calculated above that the average motion of an ensemble of particles is 0. This does not mean, however, that *every* particle remains at the origin. Rather, it simply implies that for every particle that has moved a given net distance to the right, there is likely to be a particle that has moved the same distance to the left. Although both particles have moved some net distance, the average of their motion is zero. This effect is shown in figure 5.1, in which five particles are tracked as each takes five hundred random steps on the x-axis.

In this respect, the average motion of particles can be a misleading index of particle transport. In many cases we are primarily concerned with how far particles have moved from their starting position, and we do not care if this motion is to the left or the right. In other words, we often care more about the *spread* of particles than their average position. On the basis of what we learned in chapter 4, this immediately leads us to another useful way of calculating particle transport.

In chapter 4, we found that the spread of a sample of random variables could be described by its *variance*. It seems appropriate, then, to use the variance of particle locations as an index of their average spread away from the origin. To calculate the variance, we follow the logic suggested by our calculation of the mean particle location.

Recall that the variance is the average squared deviation from the mean. In the case of particles whose initial location was at the origin ($x = 0$), we now know that the mean location after any number of steps is still at the origin (eq. 5.11). In this special case, then, the location of a particle on the x-axis and its deviation from the mean location are the same thing. Thus, to calculate the variance of particles' locations, we simply need to calculate the average square of their locations after n steps. As before, we do this by expressing the location of a particle after n steps in terms of its location after $n - 1$ steps. Expressed as an equation:

$$x_i^2(n) = [x_i(n - 1) + \delta]^2. \tag{5.13}$$

The square of the location of the ith particle after n steps is equal to the square of [the location after $n - 1$ steps plus the length of one step]. As before, δ can be positive or negative with equal probability. Squaring the term in brackets,[2] we see that

$$x_i^2(n) = x_i^2(n - 1) + 2x_i(n - 1)\delta + \delta^2. \tag{5.14}$$

We now proceed to average over the N particles in our ensemble:

$$\overline{X^2}(n) = \frac{1}{N}\sum_{i=1}^{N}[x_i^2(n - 1) + 2x_i(n - 1)\delta + \delta^2],$$

$$= \frac{1}{N}\sum_{i=1}^{N}x_i^2(n - 1) + \frac{1}{N}\sum_{i=1}^{N}2x_i(n - 1)\delta + \frac{1}{N}\sum_{i=1}^{N}\delta^2. \tag{5.15}$$

The second of these summations is similar to the one we encountered when calculating the average particle position. Because steps are taken randomly to

[2] At this point people often get confused with the notation. Recall that $x_i(n - 1)$ is the location on the x-axis of the ith particle after $n-1$ steps. The square of this value is $x_i^2(n-1)$, *not* $x_i^2(n - 1)^2$. That is, we are squaring the magnitude of the location but still want that value at $n - 1$ steps from the start of our random walk.

the right and left, we expect the sum of $2x_i(n - 1)\delta$ to be 0 across many particles, and this term thus drops out of consideration. As a result, the mean square location of a particle (its variance) is

$$\sigma_X^2(n) = \overline{X^2}(n) = \frac{1}{N} \sum_{i=1}^{N} x_i^2(n - 1) + \frac{1}{N} \sum_{i=1}^{N} \delta^2,$$
$$= \overline{X^2}(n - 1) + \delta^2. \tag{5.16}$$

In other words, the variance in particle location after n steps is equal to the variance after $n - 1$ steps plus the square of the step length. Given that all particles are initially at the origin, so that $\sigma^2(0) = 0$, we can conclude that the variance after one step is δ^2. After two steps, the variance is $2\delta^2$, and after n steps the variance must be $n\delta^2$. This, then, is our answer: Given a sufficiently large ensemble (that is, when N is very large),

$$\sigma_X^2(n) = \overline{X^2}(n) = n\delta^2. \tag{5.17}$$

The variance of particle location is equal to the product of the number of steps and the square of step length.

It is often useful to take the square root of the variance so as to express the spread of particles in units of distance rather than distance squared. Thus, the spread of particles undergoing a random walk along the x-axis is commonly expressed as the root mean square location (equivalent to the standard deviation):

$$\sigma_X(n) = X_{\text{rms}}(n) = \sqrt{\overline{X^2}(n)} = \sqrt{n}\delta. \tag{5.18}$$

To give this conclusion some tangibility, let us take the calculation one step further and express n, the number of steps the particle takes, in terms of time, t. From the rules of our random walk, we know that one step is taken every τ seconds. Thus,

$$n = \frac{t}{\tau}, \tag{5.19}$$

and inserting this value for n into eq. (5.18), we see that

$$\sigma_X(n) = X_{\text{rms}}(t) = \sqrt{\frac{t}{\tau}}\delta \tag{5.20}$$

5.2.3 DIFFUSIVE SPEED

Unlike the conclusion we reached regarding the mean location, this fact regarding the spread should be somewhat surprising. What this equation tells us is that, on average, the distance a particle moves away from its starting position increases linearly with the *square root* of time. This is in contrast to more familiar types of motion. For instance, if you are driving a car, your average speed is equal to the distance you have traveled divided by the time you have been driving:

$$u = \frac{x}{t}. \tag{5.21}$$

If x increases in linear proportion to t, u is constant. By analogy, we can calculate an average "speed" for our randomly walking particle by dividing x_{rms} by time:

$$u_{rms} = \sqrt{\frac{1}{\tau t}}\delta. \tag{5.22}$$

Unlike x in eq. (5.21), δ and τ do not increase through time, and as a result speed *decreases* as time goes on.[3] To mimic in a car the transport associated with a random walk, you would have to ease up on the throttle so that 4 seconds after you start the experiment your speed has decreased by half. After 16 seconds it is decreased by half again, and by half again at 64 seconds. This strange property of a random walk has important consequences.

5.3 *Diffusion and the Real World*

At this point it is useful to introduce an important definition. We define a *diffusion coefficient, D*:

$$D \equiv \frac{1}{2}\frac{d\sigma^2}{dt}. \tag{5.23}$$

Expressed in words: the diffusion coefficient is equal to half the rate at which the variance of particle location changes through time.[4] For our one-dimensional

---

[3] Conversely, speed increases as t gets small and would approach infinity if t could go to zero. But t is the time during which a particle has been performing a random walk. For $t \leq \tau$, the particle is in the middle of a step, and therefore is not moving randomly. Thus, τ is the minimum for t, and sets the maximum u_{rms}.

[4] The reason for the factor of 1/2 will become apparent when we discuss Fick's law later in this chapter.

TABLE 5.1 The Diffusion of Small Molecules in Water

X_{rms}	Time (s)	Time (yr)
1 μm	$5 \cdot 10^{-4}$	
1 mm	500	
1 cm	$5 \cdot 10^{4}$	
1 m	$5 \cdot 10^{8}$	15.85
10 m	$5 \cdot 10^{10}$	1585
100 m	$5 \cdot 10^{12}$	158,500
1 km	$5 \cdot 10^{14}$	1,585,000

Note: The time it takes to travel a given distance increases as the square of the distance.

random walk, $\sigma_X^2 = \delta^2 t / \tau$, so that in this case

$$D = \frac{\delta^2}{2\tau}. \tag{5.24}$$

Rewriting eq. (5.20) in terms of this diffusion coefficient, we see that

$$\sigma_X(t) = X_{rms}(t) = \sqrt{2Dt}. \tag{5.25}$$

So, if we can deduce or measure D, we can calculate the spread (expressed as the standard deviation, X_{rms}) as a function of time.

We are now in a position to relate our mathematics directly to the real world. The diffusion coefficient is a function of temperature, the mass and shape of an object, and the fluid in which it moves, and physical chemists have empirically measured D for a wide variety of molecules. The diffusion coefficients in water for small molecules (such as sucrose) are on the order of 10^{-9} m^2 s^{-1}. Given this value, we can predict how far (on average) an aquatic molecule wanders away from its starting position in a given period. The results are shown in table 5.1. Even though at 0°C a sucrose molecule moves at an average speed of 81 m s^{-1}, because its direction is randomly changing, it takes an average of 500 seconds to move an rms distance of a millimeter and nearly 14 hours to move a centimeter. This, then, is why you have to stir your iced tea if you want the sugar well mixed. If you were to rely on diffusion alone, you would have to wait years for the sugar to reach the top of your glass in tastable abundance.

An even more extreme example can be found in the sun. A photon of light produced at the center of the sun in effect diffuses its way to the surface, continually being scattered by its interactions with particles of matter. Whereas a photon, moving at the speed of light, would take only a bit more than 2 seconds to travel in a straight line from the center of the sun to its periphery (a distance of 695,950 km), it in fact takes *30,000 years* to make the journey, the increased time being due to the convoluted path the photon takes in its interactions with matter (Shu 1982).

5.4 *A Digression on the Binomial Theorem*

Before we explore the uses of the theory of molecular diffusion, let's digress briefly to examine in greater detail the connection between the theory of probability distributions and the random walk. We begin by noting that a one-dimensional random walk is simply one form of a binomial experiment, the sum of repeated Bernoulli trials. Every time one of our particles takes a step, there are two mutually exclusive outcomes. Either the step is to the right (in the positive x direction) or to the left (in the negative x direction). In analogy to chapter 4, we could arbitrarily define a step to the right as a "success" and a step to the left as a "failure." The location of the particle along the x-axis then depends on the number of trials (= steps) and the relative number of successes and failures in this binomial process. Furthermore, the probability of these relative numbers is something we have learned to calculate using the binomial theorem.

Consider an example. If ten steps are taken at random, what is the probability that exactly five of those are to the right and five to the left? That is, what is the probability that a particle undergoing a random walk of ten steps will return precisely to its starting point? From the binomial theorem, we know that the answer to this question is

$$P(x = 0) = \binom{n}{n/2} p^{n/2} q^{n-n/2} = \binom{10}{5} p^5 q^5. \tag{5.26}$$

(Note that if an odd number of steps is taken, given our rules of a random walk it is impossible for a particle to return to its starting point. Thus, in this formula n must be an even number.) For our random walk we know that $p = q = 1/2$. Working through the math, we see that the probability after ten steps that a particle will be back at the origin is only 0.246. After twenty steps, the probability is 0.176, after fifty, 0.112. In other words, our knowledge of probability leads us to expect that it is unlikely that a particle undergoing a random walk will return *precisely* to its starting point, and the probability of returning to the origin decreases as time goes on. We should indeed expect particles gradually to spread out as seen in figure 5.1.

In what pattern do the particles spread? Again, our knowledge of the binomial theorem provides an easy answer. Recall that for large n, the binomial distribution is approximated by the normal distribution. The normal distribution, in turn, is specified by its mean and standard deviation. But we already know the mean and standard deviation for a group of particles randomly walking along an axis. The mean μ_X is just the position at which the particles were released, in our example $x = 0$. From eq. (5.25) we know that the standard deviation $\sigma_X(t) = \sqrt{2Dt}$. We can thus describe the diffusive spread of a large number of particles by writing the equation for the normal probability density function using these terms:

$$g(x) = \frac{1}{\sqrt{4\pi Dt}} \exp\left(-\frac{x^2}{4Dt}\right). \tag{5.27}$$

The probability that at time t, $a \leq x \leq b$ is just the integral of this expression between a and b. Neat!

We can put our knowledge of probability theory to even further use. In taking a step, each of our randomly walking particles moves either to the right or left with equal probability, a fact that allowed us to conclude that the expected position of a particle does not change through time. But the expectation is a value based on a very large number of samples. In any finite ensemble of N particles, how far from the starting position is the actual average likely to deviate? As we have seen in chapter 4, the variability of the mean is aptly described by the standard error, so to answer this question we calculate the standard error for our random walk.

Recall that the standard error is the square root of [the variance of the distribution from which observations are taken, divided by the number of observations]. In this case we know the number of observations (N, the number of particles taking steps), so our remaining task is to calculate the variance of the stepping process. Simply done:

$$\begin{aligned} \text{variance} &\equiv E[X - \mu_X]^2 \\ &= p \cdot (\delta - 0)^2 + q \cdot (-\delta - 0)^2. \end{aligned} \tag{5.28}$$

Given that $p = q = 1/2$, the variance associated with one step is

$$\sigma_X^2(1) = \delta^2, \tag{5.29}$$

the same conclusion we arrived at earlier by a different route. The variance associated with the sum of n independent steps is simply the sum of individual variances,

$$\sigma_X^2(n) = n\delta^2, \tag{5.30}$$

again the result we calculated earlier. Thus,

$$\text{standard error} = \sqrt{\frac{n\delta^2}{N}} = \frac{\sqrt{n}\delta}{\sqrt{N}}. \tag{5.31}$$

The larger the step length (δ) and the more steps that are taken (n), the larger the error we are likely to make in assuming that particles, on average, go nowhere. Conversely, however, the larger the number of particles we use in our ensemble, the smaller the error. For real-world examples of molecular diffusion, we seldom deal with ensembles that contain fewer than a billion molecules (10^{-14} moles), so the practical error is negligible in assuming that the mean position remains constant. In cases other than those involving molecules, however, small ensembles are commonly encountered, and the error we have calculated here may need some attention. We will return to this point when we explore genetic drift in chapter 6.

5.5 *The Biology of Diffusion*

We now turn our attention to the biological consequences of molecular diffusion. Let's start by putting into a biological context the speed at which molecules are transported by diffusive random walks. We can accomplish this (and have some fun at the same time) by holding a race. Imagine the following hypothetical situation. We position a motile organism and a small, compact cloud of diffusible molecules at the starting gate and release them simultaneously. The organism swims in a given direction at a constant speed, and the molecules diffuse. It's a race to see which can move farther in a given period of time, but a race with some very strange properties.

First, we know who will eventually win. As we saw earlier, the diffusive "speed" of the cloud of molecules is analogous to a car that is continually slowing down. If the rate of spread of the molecular cloud is constantly decreasing while the speed of the organism stays the same, the organism must eventually move ahead of the molecules (fig. 5.2). The suspense in the race, then, is in figuring out how long it will take the goal-oriented animal to take the lead. To arrive at the answer, we could equate the distance traveled by the participants:

$$u_0 t = \sqrt{2Dt}, \tag{5.32}$$

where u_0 is the constant speed of the organism and $\sqrt{2Dt}$ is the root mean square distance traveled by the molecules due to diffusion.

Recall, however, that the root mean square is (in this case) the same as the standard deviation of distance traveled. For a normal distribution of distances (and as we have seen, the distances traveled by randomly walking particles will be approximately normally distributed), 68% of particles will have traveled less than one standard deviation. This biases the race a bit; we are requiring our goal-oriented animal to outstrip all but 32% of the diffusing molecules. If, however, we use $(2/3)\sqrt{2Dt}$ as our distance traveled by diffusion, we have the distance

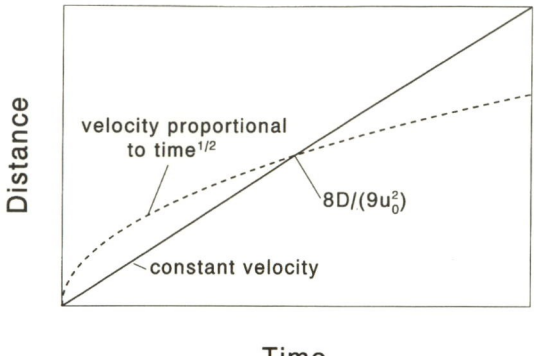

velocity proportional
to time$^{1/2}$

$8D/(9u_0^2)$

constant velocity

Time

FIG. 5.2 Because the "speed" of diffusion continually slows through time (the dashed line), an organism moving with a constant velocity (the solid line) eventually wins the race described in the text.

traveled by approximately 50% of the particles. Thus, to be fair, let's define our race by setting

$$u_0 t = \frac{2\sqrt{2Dt}}{3}. \tag{5.33}$$

Solving for t, we find that

$$t = \frac{8D}{9u_0^2}. \tag{5.34}$$

This is the answer we seek, the time at which the organism pulls ahead of most of the diffusing particles.

We can take this calculation one step further. Many small aquatic animals, from the size of bacteria to trout, can sustain a maximum speed of about ten body lengths per second (Denny 1993). Given this estimate of speed, we can restate eq. (5.34) in terms of body length, L:

$$t = \frac{8D}{9(10L)^2} \cong \frac{0.009D}{L^2}. \tag{5.35}$$

So, the larger the organism, the shorter the time it takes to win the race.

A few examples are in order. Let's assume that the molecular competitor in our race has a diffusion coefficient of 10^{-9} m^2 s^{-1} (a reasonable value for small molecules), and we will compare its motion to organisms of various sizes. Really small organisms (viruses and rickettsia, for instance) aren't motile (for reasons we will soon be able to guess), but we can explore the consequences of their hypothetical mobility by resorting to science fiction. In a classic example of Hollywood science, the cast of the film *The Fantastic Voyage* (1966) is placed in a submarine, reduced tremendously in size, injected into the blood stream of a human being, and asked to cruise to the brain to correct some malady. The final size of the sub is never fully revealed (indeed it seems to vary at the demand of the plot), but there is a time when it is attacked by antibodies that are a sizable fraction of the sub's length, which makes it pretty small. Just for fun, let's let our science fiction sub be 0.1 μm long (about fifty times the size of an antibody), and we will assume that (as with an organism) it is capable of traveling ten "body" lengths per second, that is, at a speed of 1 μm per second. What would happen if it were entered in our race? Working through the math, we see that it will take 900 seconds for the tiny sub to pull ahead of the bulk of the diffusing molecules. Now, 900 seconds is 15 minutes, about a sixth of the entire movie! At the small size demanded by the plot, the fictional adventurers in the movie could perhaps have saved on energy costs by shutting down the sub's propulsion unit for the initial part of the voyage, choosing instead to let the sub simply diffuse its way toward its goal.[5] Perhaps, too, this is a reason why motility has not evolved in viruses and rickettsia.

At the next size up, let's use a real organism. A bacterium such as the *E. coli* found in our guts has a length of about 2 μm and can travel at 20 μm per second. In a race between molecules and bacteria, it takes only about 2.25 seconds for the bacteria to win. A small increase in size substantially increases the relative value of directed transport.

Copepods are a reasonable choice for the next larger size. These ubiquitous crustacean swimmers of oceans and lakes are typically one millimeter long, and it would take them only 9×10^{-6} seconds to win our race. At the next size up, an anchovy 10 centimeters long would win in 9×10^{-10} seconds, essentially instantly. The message here is that except at the spatial scale of molecules, diffusion cannot hold a candle to directed transport.

There are cases, however, where molecular scales are not the stuff of science fiction and are indeed the biological norm. In these cases, diffusion is second to none. Consider again the bacterium we dealt with above. We calculated that it would take the organism only 2.25 seconds to outrun most of the diffusive cloud. True enough, but at a speed of ten body lengths per second, this means that the bacterium must move the considerable distance (relative to its diminutive size) of 22.5 body lengths before it pulls ahead. Over any shorter distance, diffusion is doing a more effective job of transport than the directed motion of the organism.

This raises questions of why and when the bacterium should bother moving. For example, if the bacterium is less than 22.5 body lengths (45 μm) away from a diffusible source of food, it will take less time for the food to come to it by diffusion than it would for it to swim to the food. In such a case it is probably not worth the energy spent in locomotion, and the bacterium (like

---

[5] There would be a risk in relying on the diffusive transport of a single sub. It is just as likely that the vessel would diffuse away from the brain as toward it. Shutting down the propulsion system would be an effective strategy only when dealing with a whole fleet of subs.

a rickettsia) can reasonably decide simply to sit and wait for dinner to arrive. Edward Purcell (1977) has likened this situation to a cow standing in a pasture in which the grass grows faster than the cow can eat it. Only if the cow decides it likes clover better than grass, and if clover is found only in the field down the road, is it worth moving.

In a related example, Berg and Purcell (1977) have calculated that a spherical cell of radius r would have to swim at a speed $2.5D/r$ to double the rate at which it encounters molecules in the surrounding medium. For a bacterium 2×10^{-6} meters long ($r = 10^{-6}$ m) and a diffusion coefficient of 10^{-9} m^2 s^{-1}, this is a speed of 2,500 μm s^{-1}. At 1250 body lengths per second, this is well beyond the capabilities of the average bacterium. The fastest bacteria known swim about ten times as fast as *E. coli* (speeds up to 407 μm s^{-1} have been recorded by Magariyama et al. 1994 and Mitchell et al. 1995a,b), but these marine bacteria are quite small ($r \cong 0.3$ μm), and they would have to swim at 8,333 μm s^{-1} to double the rate at which they encounter molecules. In other words, the rate of diffusive delivery of molecules is so high at the scale of a bacterium that extraordinary (and probably impossible) efforts would have to be made to increase delivery by very much.

An even more convincing example of the utility of diffusion is found in the junction between vertebrate motor nerves and muscles. The function of the junction is to allow an electrical action potential traveling down the nerve cell to trigger the contraction of the muscle. But the signal is passed from nerve to muscle not as an electrical current. Instead, special vesicles in the nerve's end plate release a neurotransmitter molecule (acetylcholine), and this molecule is transported by diffusion across the gap between cells, a distance of about 50×10^{-9} m. When they arrive at the cell membrane of the muscle, the acetylcholine molecules bind to receptors that initiate the process of muscular contraction. If diffusion were slow, the transmission of information from nerve to muscle would be sluggish, and muscular control would be severely constrained. How long do we expect it will take for an acetylcholine molecule to travel 50 nm? Setting X_{rms} to 50×10^{-9} in eq. (5.25), and D to 10^{-9} m^2 s^{-1}, we calculate that about 32% of the acetylcholine molecules will cross the gap in 1.25 *micro*seconds or less.[6] Thus, for the small scale of a neuromuscular junction, directed transport is totally unnecessary; diffusion alone can handle the job more than adequately.

---

[6] This calculation tacitly assumes that molecules are prohibited by the presence of the nerve endplate from diffusing away from the muscle. In effect, we assume that molecules are "reflected" from the surface of the nerve, allowing us to count both tails of the normal distribution. Diffusion in the presence of this type of reflecting boundary will be treated in greater depth in chapter 6, when we discuss diffusion to capture.

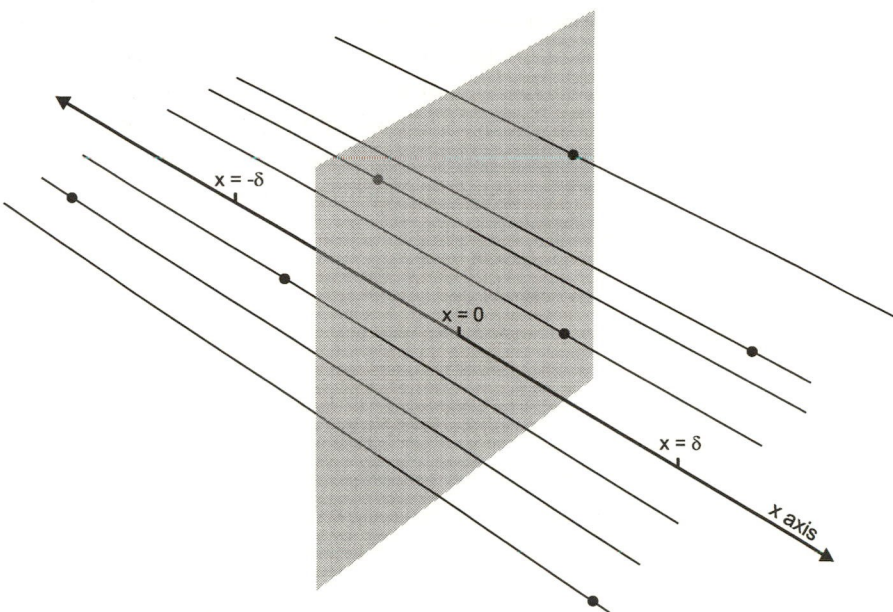

FIG. 5.3 A schematic diagram of the situation used to derive Fick's first equation of diffusion. Each particle wanders along an axis parallel to the x-axis and is free to wander back and forth through area A, shown as the shaded square.

5.6 *Fick's Equation*

So far we have examined the process of diffusion by following the motion of individual particles. There are many cases, however, in which it is advantageous to translate this microscopic viewpoint into a macroscopic perspective. This translation involves a shift from the mathematics of probability to that of differential equations, a process that is instructive to follow.

Consider the situation shown in figure 5.3. We again examine the case of a one-dimensional random walk, specifically a walk parallel to the x-axis. In this case, rather than dealing with multiple particles on a single axis, we suppose that we have multiple parallel axes, each with a single particle. We erect a hypothetical area A perpendicular to the axes, located at $x = 0$. Particles can move through this area as they step back and forth, and we desire to keep track of the *flux* of particles, that is, the net number of particles moving per time to the right through A.

To calculate the flux, we again appeal to the rules of the random walk. As before, each particle takes a step of length δ every τ seconds. This simplifies matters a bit. Only those particles that lie within a distance δ of area A can cross through the area with their next step, so we need only take into consideration those particles that initially lie between $-\delta$ and $+\delta$. If at time t there are $N(l)$ particles to the left of area A (that is, their starting position is between $-\delta$ and 0), by time $t + \tau$ half of these particles will have stepped to the right and thus will have crossed through our hypothetical area. If at time t there are $N(r)$ particles to the right of area A, we expect that at time $t + \tau$ half these particles will have stepped to the left through our area. Thus, the *net* number of particles moving to the right through A is

$$\text{net number moving to the right} = \frac{1}{2}N(l) - \frac{1}{2}N(r). \qquad (5.36)$$

Rearranging, we see that

$$\text{net number moving to the right} = -\frac{1}{2}[N(r) - N(l)]. \qquad (5.37)$$

Now, flux is defined as the net number per time per area, so to calculate the flux (symbolized by J_x) we divide by the time over which the movement of particles occurred (τ) and the area through which they moved (A):

$$J_x = -\frac{1}{2}\frac{[N(r) - N(l)]}{A\tau}. \qquad (5.38)$$

Now for some mathematical legerdemain. We multiply the right side of eq. (5.38) by 1 in the guise of δ^2/δ^2 and rearrange:

$$J_x = -\frac{\delta^2}{2\tau}\frac{1}{\delta}\left[\frac{N(r)}{A\delta} - \frac{N(l)}{A\delta}\right]. \qquad (5.39)$$

We then note that the product of A (an area) and δ (a length) is a volume. This means that the quantity $N(r)/(A\delta)$ is the number of particles per volume, or a *concentration*. Specifically, it is the concentration of particles immediately to the right of area A. For purposes that will become clear in a moment, it will be useful to assign a particular location to this concentration. Recall that $N(r)$ is the number of particles that lie anywhere between 0 and δ to the right of area A. If the particles are uniformly distributed, their average position is $\delta/2$, and we can use this as a characteristic location. Similarly, $N(l)/(A\delta)$ is the concentration of particles to the left of A, at a characteristic location $-\delta/2$. If we use the symbol C to signify concentration, we can then rewrite eq. (5.39) as

$$J_x = -\frac{\delta^2}{2\tau}\frac{[C(\delta/2) - C(-\delta/2)]}{\delta}. \qquad (5.40)$$

A further simplification is possible when we recall that the term $\delta^2/(2\tau)$ is the diffusion coefficient for our one-dimensional random walk (eq. 5.24).[7] Thus, the

---

[7] The factor of 1/2 in the definition of the diffusion coefficient can thus be seen to arise from the step probability we have assumed in our random walk.

result of our mathematical machinations is that

$$J_x = -D\frac{[C(\delta/2) - C(-\delta/2)]}{\delta}. \tag{5.41}$$

Although we assume throughout this text that you have had a course in introductory calculus, we also assume that you (like most of us) have forgotten some of the finer points of the derivations. To refresh your memory, we ask you to harken back to your days in freshman calculus, and to recall the following exercise. Take the difference between the values of a function at two points on the x-axis. If you let those points get closer and closer together, the difference between the values of the function becomes smaller and smaller and tends to zero. But if you divide this difference by the distance between points, as the points get closer and closer, the *ratio* of difference to distance may tend toward a nonzero value. In fact, in the limit as the separation between points approaches zero on the x-axis, the ratio of difference to distance defines the slope or *gradient* of the function. Ring a bell? If so, a careful perusal of eq. (5.41) reveals an interesting point. By definition:[8]

the gradient of concentration, $\dfrac{dC}{dx} = \dfrac{[C(\delta/2) - C(-\delta/2)]}{\delta}$ as $\delta \to 0$. (5.42)

Thus, our final simplification:

$$J_x = -D\frac{dC}{dx}. \tag{5.43}$$

The flux of particles (net movement per time per area) is proportional to the gradient in concentration, and the proportionality constant is nothing other than the diffusion coefficient. The negative sign tells us that the direction of net transport is from areas of high concentration to those of low concentration. This is *Fick's first equation of diffusion*, and it is the entry into the practical use of diffusion theory in much of biology, physics, and engineering.[9]

Before we move on to the uses of Fick's equation, it is useful to note that the same logic we applied here to calculate the flux along the x-axis can be

[8] In this derivation, we must remember that δ is a small, but finite, length—the distance a particle moves in a single step. As a result, we are only approximating the derivative here, using a finite difference to estimate an infinitesimal difference. The approximation is a very good one, however, and we will treat it as if it were exact.

[9] There is a second equation of diffusion attributable to Fick; it describes the temporal rate of change of concentration. The examples we discuss in this chapter assume that the rate of change of concentration is zero, and we will therefore not have occasion to use Fick's second equation.

applied in other directions as well. Thus, we can write analogous equations for the fluxes along the y- and z-axes:[10]

$$J_y = -D\frac{dC}{dy}, \tag{5.44}$$

$$J_z = -D\frac{dC}{dz}, \tag{5.45}$$

and even for the radial flux toward or away from a central point:

$$J_r = -D\frac{dC}{dr}. \tag{5.46}$$

In this case, r is the radial distance from the point of symmetry.

5.7 *A Use of Fick's Equation: Limits to Size*

One of the primary advantages of working with the macroscopic viewpoint of diffusion (in the guise of Fick's equation) is that we can immediately tap into a sizable reservoir of existing mathematics. Over the years, mathematicians, physicists, and biologists have put forth considerable effort to solve eqs. (5.43)–(5.46) under various conditions, and we are free to use these results. Consider, for example, a problem common to biology at the spatial scale of individual cells. A sphere of radius r_c (a reasonable model of a cell) is immersed in a fluid with a concentration C_∞ of a particular molecule (sucrose or oxygen, to use familiar examples). We assume that the volume of fluid is so large that this concentration of sucrose or oxygen is maintained in the bulk of the fluid regardless of any fluxes in the immediate vicinity of the cell. We now ask the question, at what rate can these metabolites be delivered to the cell if their transport relies on diffusion alone?

To answer this question we make a simplifying assumption: we suppose that the cell is a perfect absorber of sucrose or oxygen. In other words, every molecule that arrives at the cell's surface is immediately immobilized and

---

[10] The fact that particles can simultaneously diffuse along several axes raises a technical matter concerning the equations for flux presented here. To be thoroughly correct, each expression for a gradient in concentration should be written as a partial derivative: $\partial C/\partial x$, $\partial C/\partial y$, $\partial C/\delta z$. In other words, although concentration could vary with respect to other factors, we are only taking its derivative with respect to the x-, y-, or z-axes, respectively. In this text, we never consider diffusion along more than one axis at a time, and within this context the use of ordinary derivatives should not be misleading. We thought it best to avoid complicating matters and have refrained from introducing new symbols that, while technically correct, have little contextual value.

engulfed, with the result that the concentration at $r = r_c$ is 0. Given this condition, it can be shown (see Berg 1983) that eventually a steady state is reached in which the concentration at radius r $(\geq r_c)$ is

$$C(r) = C_\infty\left(1 - \frac{r_c}{r}\right). \tag{5.47}$$

Because the sphere is radially symmetrical, this spatial variation in concentration depends only on distance from the cell; it is independent of direction. Taking the derivative of this expression with respect to r and inserting the result into the radial version of Fick's equation (eq. 5.46), we find that

$$J_r(r) = -DC_\infty\frac{r_c}{r^2}, \tag{5.48}$$

and at the cell's surface $(r = r_c)$,

$$J_r(r_c) = -\frac{DC_\infty}{r_c}. \tag{5.49}$$

In other words, the flux of metabolites to the cell is radially inward (the negative sign tells us this) and is inversely proportional to the cell's radius, r_c. The larger the diffusion coefficient and the higher the bulk concentration of diffusing molecules, the faster the flux (as you might expect), but the larger the cell, the slower the flux (as you may well not expect). Why should the flux depend on the size of the cell?

The answer lies in the radial symmetry of the situation. Imagine an infinitesimally small sphere that is absorbing molecules at a constant rate of Q particles per second (fig. 5.4). Molecules arrive at this "sink" equally from all directions, and we desire to keep track of how they are flowing. To do so, we mentally erect a permeable, spherical shell around the sink and count how many molecules per time pass through it on their way inward. If the flow is constant (and we have assumed that it is), Q molecules must pass through our shell each second to make up for the Q molecules absorbed by the sink. In other words, to keep the *current* (molecules per second) flowing at a constant rate, the flow must be the same through the shell as it is to the sink. Note, however, that we have not defined the size of our shell. In fact, its size does not matter; any shell will experience the same current regardless of its radius. Furthermore, as molecules move steadily inward, we can imagine them passing through a whole series of permeable shells (like the layers in an onion), with the same current crossing each shell.

Now, remember that flux is the net number of particles per time passing through an area. So, we can calculate the flux through any of our onion-like

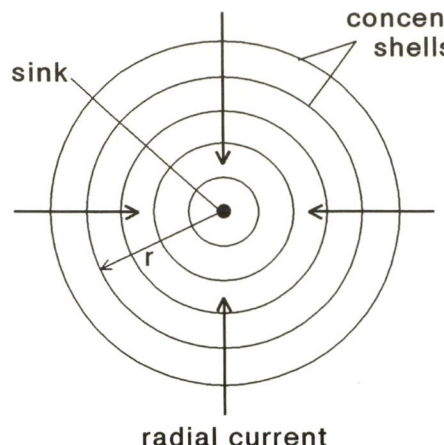

sink

concentric shells

radial current

FIG. 5.4 As particles diffuse radially inward toward a central "sink," we can think of them as flowing through a series of permeable, concentric, spherical shells (here shown in cross section). The situation depicted here can be used to explain why flux (current per area) varies with radial distance (r) from an absorbing sphere.

shells by dividing Q by that shell's area. Note, however, that the area of one of our spherical shells *does* depend on its size:

$$\text{surface area of a sphere} = 4\pi r^2. \tag{5.50}$$

Thus, if the current is constant, the flux (current per area) must be

$$\text{flux} = \frac{Q}{4\pi r^2}. \tag{5.51}$$

The larger the shell, the smaller the flux because larger shells have larger areas. In essence, the membrane of a cell acts like a spherical shell, and this leads to the size dependence of the flux noted in eq. (5.49).

Now, back to our metabolizing cell, where we turn this argument around. If we want to know the total rate at which sugar or oxygen is being delivered to the cell (rather than the rate per area we calculated in eq. 5.49), we need to multiply the flux by the surface area of the cell. The surface area of a sphere of radius r_c is $4\pi r_c^2$, so the inward *current* (the net number of particles per second across the entire cell area) is

$$I = |-4\pi DC_\infty r_c|. \tag{5.52}$$

When you think about it, this conclusion is a bit strange. Even though the area through which a cell absorbs nutrients increases with the square of radius, the inward current of molecules increases only with radius to the first power.

This disproportionate scaling can have important results. Let's see what its consequences are if our model cell (like real cells) requires that metabolites be delivered at some minimal rate if the cell is to survive. We will assume as a first approximation that the metabolic needs of the cell are proportional to the amount of cytoplasm it contains, and thus to its volume.[11] The volume of a spherical cell of radius r_c is $(4/3)\pi r_c^3$, so we can set the overall metabolic requirement to

$$\text{metabolic need} = \frac{4\pi r_c^3 M}{3}, \tag{5.53}$$

where M is a metabolic rate coefficient, expressed in moles per second per cubic meter.

Because the metabolic need of a cell increases with radius to the third power, while delivery of metabolites increases only with the first power of radius, there must be an upper limit to the size of cells if they are supplied by diffusion alone. We can calculate what this limit is. To do so, we set metabolic delivery (diffusive current) equal to metabolic demand,

$$4\pi DC_\infty r_c = \frac{4\pi r_c^3 M}{3}, \tag{5.54}$$

and solve for r_c. Thus, the maximal radius of a cell supplied solely by diffusion is

$$r_{c,\,\text{max}} = \sqrt{\frac{3 D C_\infty}{M}}. \tag{5.55}$$

The larger the diffusion coefficient and the higher the concentration of metabolites, the bigger a cell can get; the larger its metabolic needs, the smaller it must be.

Phytoplankton provide a relevant example. Many of these free-floating aquatic organisms absorb bicarbonate ions (HCO_3^-), and use them as a source of carbon for photosynthesis. Lacking any means of actively transporting either themselves or the fluid around them, phytoplankton rely on diffusion for the delivery of their carbon source. An actively photosynthesizing diatom consumes about one mole of bicarbonate per second per cubic meter of cell, the concentration of bicarbonate in typical seawater is about 1.5 moles per cubic meter, and a representative diffusion coefficient for bicarbonate in water is 1.5×10^{-9} m^2 s^{-1}. How big can the cell be and still have its carbon demand met by diffusion alone? Utilizing eq. (5.55) we find that the answer is a radius of about 80 μm.

---

[11] For many multicellular organisms, metabolic rate is proportional to volume$^{3/4}$. (More precisely, it is proportional to mass$^{3/4}$, but given a relatively constant density among organisms, the two expressions are comparable.) It is less clear that this same scaling holds for the single cells discussed here, or for photosynthesis. To keep the math simple, we assume that metabolic rate is proportional to volume^1 (that is, volume to the first power); the message is the same one way or the other.

A single-celled planktonic plant could be larger than this, but its rate of photo-synthesis would then be constrained. Indeed, 80 μm is approximately the size of many diatoms in the sea.

What about organisms in air? The diffusion coefficients of small molecules in air are about ten thousand times those in water (Denny 1993), suggesting that a one-celled aerial organism could be a hundred times larger than its aquatic cousins and still be adequately supplied by diffusion (eq. 5.55). Visions of aerial phytoplankton 8 mm in radius come to mind.

In the case of a photosynthetic cell, it is unlikely that matters are quite so skewed, however. Unlike an aquatic plant, which has HCO_3^- to draw on at a concentration of about 1.5 moles per cubic meter, an aerial plant must get its carbon from carbon dioxide, which is present in air at a concentration of only about 0.014 moles per cubic meter. This roughly 100-fold decrease in C_∞ offsets some of the 10,000-fold increase in diffusivity, and on the basis of diffusive delivery of carbon we might expect aerial phytoplankton to be 800 μm in radius.

In reality, aerial plants of this size are hard to find, but their scarcity is likely not due to the strictures of diffusion. Biological particles (such as phytoplankton) have approximately the same density as water (1000 kg m^{-3}) and are therefore much denser than air (about 1.2 kg m^{-3}). As a result, all but the most minute particles quickly settle out of the atmosphere. For instance, a spherical plant cell 800 μm in radius would have a sinking velocity of several meters per second in air (Denny 1993). Phytoplankton in water are close to being neutrally buoyant, and therefore have much slower sinking velocities.

5.8 *Receptors and Channels*

The examples above concerned diffusion through the surface of a sphere. Other geometries can be important in biology as well. Consider, for example, the rate at which molecules can be absorbed by a circular disk, one face of which is exposed to the medium. Whereas a sphere served as a convenient model for a cell, this disk can serve as a model for the ion channels and receptors found in the cell's membrane. Because much of how a cell communicates with its surroundings is controlled by the diffusive arrival of small molecules at these receptors and channels, it is important to understand how the process works.

Berg (1983) has shown that the diffusive current to an isolated circular disk is

$$I_d = 4Dr_dC_\infty, \tag{5.56}$$

where r_d is the radius of the disk.[12] This expression is quite similar to that for the diffusive current to a sphere. The rate at which molecules are delivered

[12] The derivation of this equation is an interesting example of the cross-fertilization of ideas in science. Berg and Purcell (1977) noted that the steady-state equation for diffusion to an

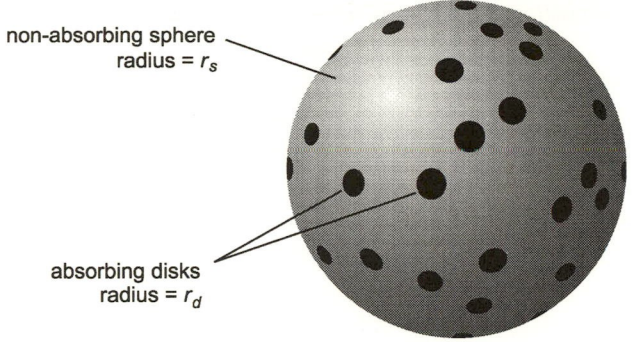

non-absorbing sphere
radius = r_s

absorbing disks
radius = r_d

FIG. 5.5 Absorbing disks (a model for receptors or ion channels) scattered on a non-absorbing sphere (a model for a cell).

increases directly with the diffusion coefficient and the bulk concentration, and with the radius (rather than the radius squared) of the disk.

This simple expression can lead to some intriguing results, however. Let's make use of our disk in its capacity as a model receptor or channel, and place it on the membrane of a spherical cell (fig. 5.5). To be precise, we assume that the sphere itself does not absorb a particular brand of small molecule that constitutes a "signal" (any signal molecule that hits the sphere merely bounces off), but that the circular disk is perfectly absorbing (any signal molecule that hits it is immediately engulfed). Thus, the presence of a disk results in a local reduction in the concentration of diffusible signal molecules (at the surface of the disk itself, the concentration is zero because all molecules are absorbed), and this concentration gradient results in a diffusive current.

So far we have not changed anything by putting our model disk on a spherical surface. The current to the disk is that described by eq. (5.56). But it would be unusual to find a cell with just a single receptor or ion channel. What happens if we allow multiple disks on our sphere? In other words, let's calculate how the current to a cell depends on the number of receptors or channels present on the cell's surface. After a bit of mathematical twisting and turning, we will find that the answer is both nonintuitive and important. In essence, the physics of diffusion allows a cell to have a variety of receptors and channels on its surface, each simultaneously acting at near-peak effectiveness. As a result, each cell

absorber is (under the appropriate conditions) analogous to Laplace's equation for electrostatic potential. This allowed them to deduce that the diffusion current to an object of a given size and shape can be calculated from knowledge of the capacitance of an isolated, conductive object of the same size and shape. A search of the literature provided data on the capacitance of an isolated, conductive disk, and voilá!

can perform several functions at once, a capability that has profound biological significance.

To start, let's be bold and find out what happens if we completely coat the sphere with disks. The exposed surface area of each disk is πr_d^2, and the overall surface area of the sphere is $4\pi r_s^2$, where r_s is the radius of the sphere. Let's let the ratio of the radii be a constant, k. Thus,

$$k = \frac{r_d}{r_s}. \tag{5.57}$$

Dividing the surface area of the sphere by the area of a single disk, we can calculate roughly how many disks it would take to cover the sphere completely:[13]

$$\text{number of disks} \cong \frac{4\pi r_s^2}{\pi (kr_s)^2} = \frac{4}{k^2}. \tag{5.58}$$

For example, if the radius of a disk is only a thousandth that of the sphere (roughly the ratio of the size of the binding site of a transport protein to the size of a bacterium), $k = 0.001$, and it takes roughly 4 million disks to coat the sphere. Let's assume that disks are cheap, so that we can easily acquire 4 million of them and coat our sphere to see what happens.

If we assume that each of these 4 million disks absorbs signal molecules as if the other disks weren't present, the total current to the sphere is simply the sum of that due to each individual disk. Thus,

$$I_{\text{total}} = 4,000,000 \times I_d = 4,000,000 \times (4Dkr_s C_\infty)$$
$$= 16,000 \times Dr_s C_\infty. \tag{5.59}$$

But wait. We know from eq. (5.52) that the current to a perfectly absorbing sphere is only $4\pi Dr_s C_\infty$, and $4\pi \ll 16,000$. Something must be wrong with our calculation of the current to a sphere coated with disks.

The mistake comes in the assumption that a disk in a group will absorb molecules as if its neighbors were not present. In fact, the presence of nearby disks changes the local concentration gradient, and closely packed disks interfere with one another. This effect can be explored through an analogous problem in the flow of water through pipes.

---

[13] No matter how you arrange them, closely packed, nonoverlapping disks always leave small bits of area uncovered. In this respect, it might be better to work with a combination of pentagonal and hexagonal disks, but then we would have to forsake the ease of computation afforded by our assumption that disks are circular. For the sake of simplicity, we ignore the complications intrinsic with packing circular disks and put up with the slight error in calculating how many it takes to cover a sphere.

To begin, we note that the diffusive current of molecules to an absorber can be thought of as the product of a potential for transporting molecules (the bulk concentration, C_∞) and the "conductance," κ, of the molecules (with units $m^3 \ s^{-1}$, a value set by the diffusion coefficient and the shape of the absorber):

$$I = \kappa C_\infty. \tag{5.60}$$

Thus, for a disk, $\kappa = 4Dr_d$, and for a sphere, $\kappa = 4\pi Dr_s$ (see eqs. (5.52), (5.56)). This relationship is analogous to the transport in pipes in which the flow of water (the current) is the product of the potential for transporting liquid (the pressure head) and the conductance of the pipes (how easily the pipes allow water to flow).

This analogy allows us to model the diffusive flow of molecules to the cell using the plumbing shown in figure 5.6. Far from the cell surface, molecules are unaffected by the interference among disks, and their conductance is high. The large-bore pipe on the right is an analogy for this high conductance; water can flow relatively freely through this "main." In contrast, once molecules are close to the cell, they are strongly affected by the interference among absorbing disks, and the conductance is consequently decreased. Here we model the absorbing disks using the group of small pipes on the left, each pipe representing a disk. Due to its diminutive bore, each of these small pipes has a low conductance. What is the overall conductance of this plumbing?

We first note that the effect of competition among disks is quite localized, and at a slight radial distance Δr away from the sphere ($\Delta r \ll r_s$), the total diffusive conductance encountered by molecules is essentially the same as it would be if the whole sphere were absorbing. In other words at $r = r_s + \Delta r$, the conductance is $4\pi D(r_s + \Delta r)$, and this is the conductance we assign to the large, single pipe.

As noted above, the overall conductance is less than this, due to the resistance of the small pipes that represent disks. By analogy, we assign each pipe a conductance $4Dr_d$, and the conductance of the N small pipes in parallel is $4NDr_d$.

All we need to do now is add the conductance of the big pipe to the combined conductances of the small pipes to arrive at the overall conductance of the system. Unfortunately, adding conductances in series isn't quite as straightforward as adding them in parallel. For example, we cannot simply add the conductance of the large pipe to the combined conductances of the small pipes ($4\pi D(r_s + \Delta r)$ plus $4DNr_d$). This would imply that if we had no small pipes at all ($N = 0$), the conductance would nonetheless be $4\pi D(r_s + \Delta r)$. In fact, with no small pipes there would be no way for water to get out of the large main, and the conductance would be zero. Analogously, with no disks, the sphere would

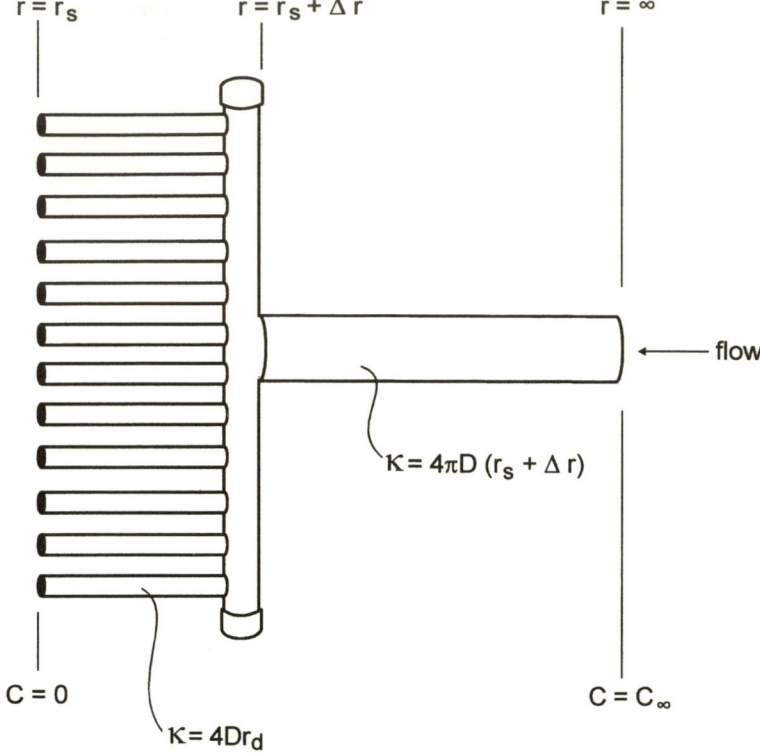

FIG. 5.6 A bit of plumbing used as an analogy to calculate the overall conductance of absorbing disks on a non-absorbing sphere. The small pipes on the left represent individual absorbing disks, and the large pipe on the right represents the bulk fluid surrounding the sphere. The analogous radii, conductances, and concentrations are noted.

be entirely nonabsorbing, and the conductance would be zero. Instead, when adding conductances, the following formula must be used:

$$\text{overall conductance} = \frac{\text{product of individual conductances}}{\text{sum of individual conductances}}. \tag{5.61}$$

As required, this expression is sensitive to small conductances, and tells us that if any single conductance in the series is zero, the overall conductance is zero.

Using this formula to add the conductance of the large pipe to that of the small pipes, we see that the total conductance is:

$$\kappa = \frac{4\pi D(r_s + \Delta r) \cdot 4NDr_d}{4\pi D(r_s + \Delta r) + 4NDr_d}. \tag{5.62}$$

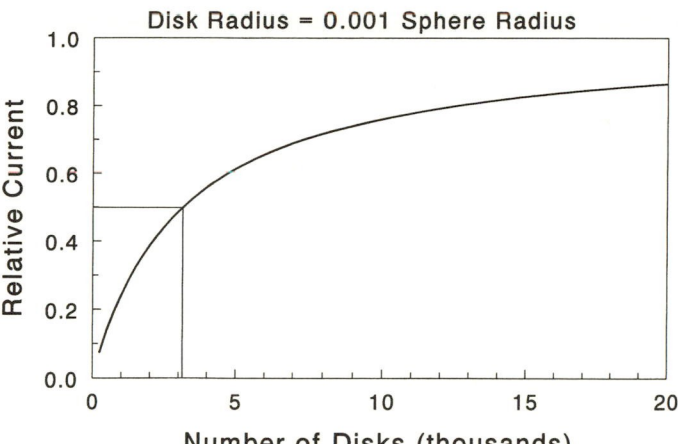

FIG. 5.7 As the number of absorbing disks increases, the relative current increases. The additional current associated with each additional disk decreases, however, so that the curve asymptotes to the current of an absorbing sphere. Half this maximal current is reached when a relatively small number of disks is present

Because Δr is small compared to r_s, $r_s + \Delta r \cong r_s$. Using this approximation and rearranging, we conclude that

$$\kappa = \frac{4\pi D r_s}{1 + [\pi r_s/(N r_d)]}. \tag{5.63}$$

Knowing the conductance, we can at last calculate the overall diffusive current to a nonabsorbing sphere with absorbing disks. From eq. (5.60):

$$I_{\text{total}} = \kappa C_\infty = \frac{4\pi D r_s C_\infty}{1 + [\pi r_s/(N r_d)]}. \tag{5.64}$$

This relationship is shown in figure 5.7.

When N is small and $r_d \ll r_s$, $I_{\text{total}} \cong 4N D r_d C_\infty$, as we would expect for small, isolated disks acting independently. However, as the number of disks grows large, they begin to interact, and the current no longer rises in linear proportion to the number of disks. The result is that the more disks we place on the sphere, the lower the current *per disk*. When N is very large, $I_{\text{total}} \to 4\pi D r_s C_\infty$, as we expect for an absorbing sphere.

How many disks would we need to add, for instance, to achieve half the maximal diffusive current? The maximum current is $4\pi D r_s C_\infty$, and if we set

I_{total} to half this value in eq. (5.64) and solve for N, we find that

$$\text{number of disks at half maximal current} = \frac{\pi r_s}{r_d}. \tag{5.65}$$

In our example, where $r_d = 0.001r_s$, 3,142 disks would be required to allow half the maximal current. But we also know that it would take approximately 4 million disks to achieve the maximal current by covering the whole sphere. Thus, half the maximal current is obtained when only $3142/4,000,000 = 0.08\%$ of the sphere's surface is taken up by disks.

We have just calculated why it is possible for a cell to perform effectively with many different types of receptors and ion channels in its membrane. If, in order to achieve a reasonable current, each type of receptor or channel had to occupy a large fraction of the cell's surface, only a few types could be present at any one time, and the cell would be limited in the number of tasks it could perform. The fact that occupation of only a very small fraction of the surface yields most of the benefits of a receptor or channel's presence means that it is feasible to have many types present at once.

Note that this conclusion is based on the fact that the diffusive flux of a given type of molecule depends on the concentration gradient of only that particular molecule. Thus, if one type of ion channel absorbs calcium, it lowers the local concentration of calcium and can compete for calcium current with other calcium channels. But—at least to a good first approximation—the local reduction in the concentration of calcium does not affect the local concentration of sodium. Dense concentrations of calcium channels may thus compete with one another without having any appreciable effect on the current moving into a nearby sodium channel.

5.9 *Summary*

Calculations regarding the allocation of membrane space among ion channels may seem a far cry from the statistics with which we started this chapter. Indeed, we have covered a lot of intellectual ground in a few pages, and it will be well to review our journey.

Random motion of molecules is an unavoidable consequence of the temperature at which we live, and the stochastic nature of molecular movement makes it impossible to predict precisely where any given molecule will be at any given time. But this random behavior on the small scale forms the basis for interesting and accurate predictions of the overall behavior of larger-scale groups of molecules. In particular, we can predict how an ensemble of molecules spreads out through time and space. These predictions, couched in terms of the binomial distribution, the normal distribution, and Fick's first equation, can be applied to a wide variety of biological processes, only a few of which we have covered here. You may wish to pursue the subject further by reading Berg and Purcell (1977), Berg (1983), or Denny (1993).

In the next chapter, we will apply the rules of random walks to objects other than molecules.

5.10 *Problems*

1. A particle undergoes a random walk along the x-axis. A step to the right (in the positive x direction) has probability p, a step to the left has probability q. Calculate and graph the probability *distribution* of this particle's location after ten steps if:

- $p = 0.5$ • $p = 0.9$ • $p = 0.3$.

2. In a particular example of a one-dimensional random walk, particles move with a step length $\delta = 1$ mm, and take steps every $\tau = 1$ s. If p, the probability of taking a step to the right is 0.6, what is the "speed" at which an ensemble of particles moves along the x-axis?

3. As we have seen, the total diffusive current to a nonabsorbing sphere depends on the number of absorbing disks on the sphere's surface. Assume a sphere of radius r_s.

- As a function of disk size (r_d), calculate the fraction of the sphere's surface that must be occupied by absorbing disks to yield half the maximal diffusive current.
- For absorbing disks of a given size, how does the current *per disk* vary as a function of the number of disks on the spheres?
- Speculate on the role of interdisk spacing in determining diffusive current. In real cells, would you expect receptors or ion channels to be spaced randomly, evenly, or in a clumped pattern?

4. The diffusion coefficient has the dimensions of *length²/time*. This may be thought of as the product of an average particle velocity (*length/time*) and a mean free path (*length*). In other words,

$$D = \sqrt{u^2}L,$$

where L is the mean free path. The diffusion coefficient for oxygen is 1.6×10^{-5} m²/s in air and 2.1×10^{-9} m²/s in water at room temperature (20° C). The molecular weight of O_2 is 5.3×10^{-26} kg. What is the mean free path of oxygen in air? In water?

5. In this chapter we contrasted the "directed" motion of a swimming bacterium with the random motion of molecules. In fact, the direction in

which bacteria swim is subject to stochastic thermal fluctuations (rotational diffusion; see Berg 1983), and the transport of real bacteria is similar in principle to a random walk. An analysis by Berg (1983) suggests that the effective diffusion coefficient of bacteria is

$$D = \frac{u^2}{6D_r},\tag{5.66}$$

where u is the bacterium's swimming speed and D_r is the rotational diffusion coefficient. For a sphere of radius r (a model for a bacterium),

$$D_r = \frac{kT}{8\pi\mu r^3}.\tag{5.67}$$

Here μ is the dynamic viscosity of water (about 10^{-3} N s m^{-2} at 20°C), k is Boltzmann's constant (1.38×10^{-23} joules per kelvin), and T is the absolute temperature. Thus,

$$D = \frac{4\pi\mu u^2 r^3}{3kT}.\tag{5.68}$$

Calculate the time required for E. coli ($u = 2 \times 10^{-5}$ m s^{-1}, $r = 10^{-6}$ m) and Vibrio harveyi ($u = 4 \times 10^{-4}$ m s^{-1}, $r = 0.3 \times 10^{-6}$ m) to travel a root mean square distance of 1 mm. Compare these results to the result of a similar calculation for small molecules in water ($D = 10^{-9}$ m^2 s^{-1}).

E. coli live in a nutrient-rich environment (the guts of mammals) while V. harveyi live in the ocean where nutrient concentrations are low and patchily distributed. Discuss the "need for speed" in the evolution of bacterial locomotion.

6

More Random Walks

In the last chapter, we explored random walks in the context of molecular diffusion, and confined our discussion to cases of motion along a single axis. But these simple examples are merely an hors d'ouevre to the rich banquet of biological problems that can be addressed using the concepts of random motion. In this chapter, we will examine two of these: the question of how long it takes an object to bump into the edge of its environment, and the question of why rubber is elastic. In the process we will expand our understanding of random walks to include motion in three dimensions and objects other than molecules (for example, the "diffusion" of alleles).

6.1 *Diffusion to Capture*

We begin with an example that is initially similar to those we explored in chapter 5, the random motion of a particle along a single axis. In this case, however, we confine the particle's motion by erecting barriers across its path. If we release the particle between these boundaries, we know that its random motions will eventually bring it into contact with one of the walls. Our job, then, is to calculate how long it will take before this inevitable collision occurs. This is the question of what Berg (1983) aptly terms "diffusion to capture"—how long a particle is likely to walk about at random in a confined space before it is captured by contact with its enclosure. Examples crop up time and again in biology: As pollen grains are wafted erratically by the wind, how long do they remain aloft before (by chance) they contact the ground? Planktonic larvae released by mussels in Nova Scotia are carried on an oceanic random walk by the turbulence of the sea; how long will it take for them to reach Ireland? In a sexually reproducing organism, the frequencies of alleles "drift" from one generation to the next in a genetic analog of diffusion; how long will it take before an allele is (by chance) lost from a population?

The essentials of these problems are embodied in the system shown in figure 6.1. Motion of the particle is along the x-axis. One of the confining barriers is erected at the origin, $x = 0$, and the other an arbitrary distance away at

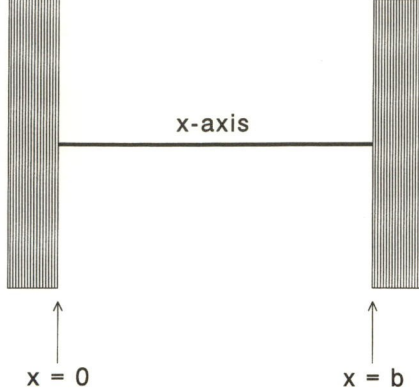

x-axis

FIG. 6.1 Diffusion to capture. The axis along which particles step is now confined between two walls. How many steps can a particle take before it hits one wall or the other?

x = 0   x = b

$x = b$. Each wall has one of two properties: it can be *absorbing*, in which case any particle that contacts the wall is immediately captured; or it can be *reflecting*, in which case a particle hitting the wall bounces back from where it came. In the examples we deal with here, the wall at the origin is always an absorbing boundary, and we explore the consequences of the wall at $x = b$ being either absorbing or reflecting.

We begin an experiment by releasing a particle on the x-axis between the barriers ($0 \leq x \leq b$) and allow the particle to move randomly along the axis according to the rules given in chapter 5: a step of length δ is taken in a random direction every τ seconds. We then ask our question: How long will we have to wait before the particle contacts an absorbing wall and is thereby captured?

It should be clear by now that there is no single answer to this question. Because the particle moves randomly, the time to capture involves a large element of chance. If we repeat the experiment ten times, we may well record ten different results. The best we can do is to calculate the *average time* it takes for a particle to blunder into an absorbing wall. We denote this average time as $\mathcal{T}$. Note that we can expect $\mathcal{T}$ to be a function of x, the position at which the particle is released. For example, if we release the particle at $x = \delta$ (that is, one step length from the absorbing boundary at the origin), there is a 50% chance that the particle will move to its left and be captured at the end of its very first step. If the particle is released 2δ from the absorbing wall, it must take at least two steps before being captured and will therefore remain free for a longer period. Thus, the time to capture is likely to be a function of the starting position. How can we calculate $\mathcal{T}(x)$?

A mechanistic approach to this problem is outlined by Berg (1983) and is somewhat nonintuitive, since we will initially take what appears to be a simple situation and make it more complex. Be assured, however, that the complexity eventually leads to a satisfyingly elegant solution.

If the particle is released at position x at time $t = 0$, at a subsequent time $t = \tau$ the particle is either at $x + \delta$ or $x - \delta$, and (according to our rules) the probability of being at either location is the same. Now, the average time it takes the particle to wander from this new position to an absorbing wall is (by definition) either $\mathcal{T}(x + \delta)$ or $\mathcal{T}(x - \delta)$, and we can restate the expected overall time to capture as

$$\mathcal{T}(x) = \tau + \frac{1}{2}[\mathcal{T}(x + \delta) + \mathcal{T}(x - \delta)]. \tag{6.1}$$

In other words, the time it takes for the particle to be captured from position x is equal to the time it takes to move one step (τ) plus the average of the times it takes to be captured from the two possible positions reached after that first step. Note that because we don't yet know the form of the function $\mathcal{T}$, we cannot assume that $\frac{1}{2}[\mathcal{T}(x+\delta)+\mathcal{T}(x-\delta)] = \mathcal{T}(x)$. In fact, this turns out not to be true.

All this may seem like nothing more than a long-winded restatement of the original proposition, but it sets the stage for some productive mathematical sleight of hand. First we move $\mathcal{T}(x)$ to the right-hand side of the equation:

$$0 = \tau + \frac{1}{2}[\mathcal{T}(x + \delta) - 2\mathcal{T}(x) + \mathcal{T}(x - \delta)]$$

$$= \tau + \frac{1}{2}\{[\mathcal{T}(x + \delta) - \mathcal{T}(x)] - [\mathcal{T}(x) - \mathcal{T}(x - \delta)]\}. \tag{6.2}$$

We then multiply both sides of the equation by $2/\delta$:

$$0 = \frac{2\tau}{\delta} + \left[\frac{\mathcal{T}(x + \delta) - \mathcal{T}(x)}{\delta} - \frac{\mathcal{T}(x) - \mathcal{T}(x - \delta)}{\delta}\right]. \tag{6.3}$$

The utility of these machinations is that the terms in the square brackets are now in the form of derivatives, as reviewed in chapter 5. For small δ,

$$\frac{\mathcal{T}(x + \delta) - \mathcal{T}(x)}{\delta} \cong \frac{d\mathcal{T}}{dx}\Big|_{x+\frac{\delta}{2}}, \tag{6.4}$$

where the subscript $|_{x+\frac{\delta}{2}}$ denotes that the derivative is taken at $x+\delta/2$. Similarly,

$$\frac{\mathcal{T}(x) - \mathcal{T}(x - \delta)}{\delta} \cong \frac{d\mathcal{T}}{dx}\Big|_{x-\frac{\delta}{2}}. \tag{6.5}$$

Although these are approximations, they are very good ones, and we will treat them as if they were exact.[1] In light of these results, we can rewrite eq. (6.3) as

$$0 = \frac{2\tau}{\delta} + \frac{d\mathcal{T}}{dx}\Big|_{x+\frac{\delta}{2}} - \frac{d\mathcal{T}}{dx}\Big|_{x-\frac{\delta}{2}}. \tag{6.6}$$

---

[1] At this point, you might be inclined to worry, "As δ gets very small, why doesn't $2\tau/\delta$ blow up?" It doesn't because δ and τ are interrelated. Recall that the average speed of a

We then divide through by δ again:

$$0 = \frac{2\tau}{\delta^2} + \frac{\frac{dT}{dx}|_{x+\frac{\delta}{2}} - \frac{dT}{dx}|_{x-\frac{\delta}{2}}}{\delta}, \tag{6.7}$$

and again apply the definition of the derivative. For small δ,

$$\frac{\frac{dT}{dx}|_{x+\frac{\delta}{2}} - \frac{dT}{dx}|_{x-\frac{\delta}{2}}}{\delta} \simeq \frac{d^2T}{dx^2}|_x. \tag{6.8}$$

Thus,

$$0 = \frac{2\tau}{\delta^2} + \frac{d^2T}{dx^2}. \tag{6.9}$$

But $2\tau/\delta^2$ is nothing other than the inverse of the diffusion coefficient (see eq. 5.24). So in one last revision we find that

$$0 = \frac{1}{D} + \frac{d^2T}{dx^2}. \tag{6.10}$$

Presto! Starting with the rules of a random walk, we have now arrived at a second-order differential equation involving only T (the function we wish to understand), x, and the diffusion coefficient, D. To the mathematically trepidatious, this may not seem like progress; but take heart: this expression is a tractable beast, and application of a bit of elementary calculus quickly takes us to our final goal.

6.1.1 Two Absorbing Walls

To solve eq. (6.10), we need to integrate twice, supplying the appropriate constant for each integration. The constants are then evaluated relative to two boundary conditions that we specify so as to tie the equation to reality. Let's assume for the moment that both the wall at $x = 0$ and at $x = b$ are absorbing. In this case, the time to capture (that is, the time until a particle is absorbed) is 0 both at $x = 0$ and at $x = b$. At these positions, a particle is already at the absorbing wall, so no further time is required to effect its capture. Thus, we specify that $T(0) = T(b) = 0$. These two equalities serve as our boundary conditions.

---

molecule is set by its mass and temperature,

$$u_{x,\,rms} = \sqrt{\overline{u_x^2}} = \sqrt{\frac{kT}{m}}$$

For a given molecule at a constant temperature, then, $u_{x,\,rms}$ is constant. Furthermore, $\delta/\tau = u_{x,\,rms}$. Thus, as δ gets small, τ must get small in like proportion.

To set up the solution, we subtract $1/D$ from each side of eq. (6.10), note that

$$\frac{d^2T}{dx^2} = \frac{d(dT/dx)}{dx}, \tag{6.11}$$

and multiply by dx:

$$d\left(\frac{dT}{dx}\right) = -\frac{1}{D}dx. \tag{6.12}$$

We then integrate:

$$\int d\left(\frac{dT}{dx}\right) = -\frac{1}{D}\int dx. \tag{6.13}$$

When the dust settles, we find that

$$\frac{dT}{dx} = -\frac{1}{D}(x + k_1), \tag{6.14}$$

where k_1 is the constant of integration. Multiplying both sides of eq. (6.14) by dx and integrating again,

$$\int dT = -\frac{1}{D}\int (x + k_1)\, dx, \tag{6.15}$$

we find that

$$T = -\frac{x^2}{2D} - \frac{x \cdot k_1}{D} + k_2, \tag{6.16}$$

where k_2 is a second constant of integration.

These two constants can now be evaluated using our boundary conditions. We know that $T(0) = 0$, from which we can conclude that k_2 must be 0. Similarly, $T(b) = 0$. Inserting b for x in eq. (6.16) with k_2 set to 0, we see that

$$0 = -\frac{b^2}{2D} - \frac{b \cdot k_1}{D}, \tag{6.17}$$

which is true only if $k_1 = -b/2$. Thus, our final answer:

$$T = -\frac{1}{2D}(x^2 - bx)$$

$$= \frac{1}{2D}(bx - x^2), \quad \text{both walls absorbing.} \tag{6.18}$$

This result is graphed in figure 6.2. Average time to capture is indeed a function of the position x at which the particle is released, rising parabolically from 0 at the walls ($x = 0, x = b$) to a maximum at $x = b/2$. This maximal time to capture is $b^2/(8D)$; that is, the maximal time to capture increases as the

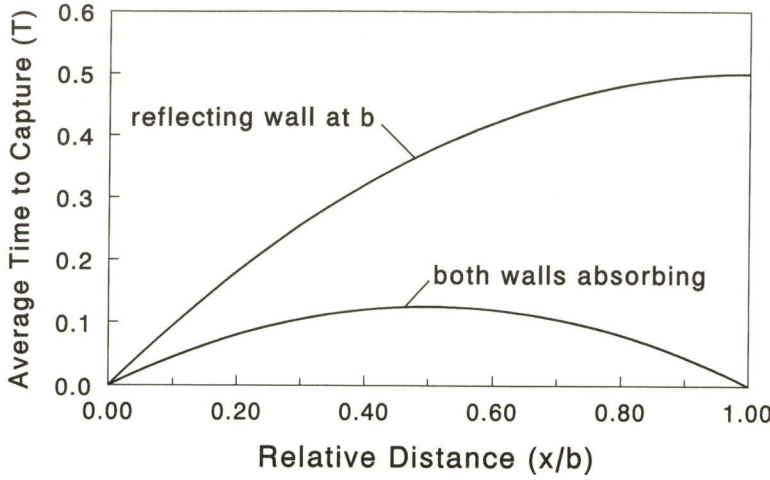

FIG. 6.2 Diffusion to capture. The average time to capture varies depending on where a particle is released and whether the walls are absorbing or reflecting.

square of the separation between absorbing walls. This should make intuitive sense if you remember our previous finding that the distance traveled by a diffusing particle increases in linear proportion to the square root of time.

There are cases in which we will not know beforehand at what position a particle begins its walk. In this situation, it can be useful to calculate an overall average time to capture for particles that have an equal probability of starting anywhere on the axis (that is, a uniform distribution) between 0 and b. This is just the formula for the expected value of a continuous random variable, as described in chapter 4. Each position x has a value $\mathcal{T}(x)$ and the probability density function for a uniform distribution (i.e., $1/b$):

$$\text{overall average } \mathcal{T} = \frac{1}{b} \int_0^b \mathcal{T}(x)\, dx$$

$$= \frac{1}{2Db} \int_0^b \left(bx - x^2\right) dx$$

$$= \frac{b^2}{12D}. \tag{6.19}$$

The overall average time to capture is 67% of the maximum ($b^2/8D$).

6.1.2 ONE REFLECTING WALL

What if one of the walls is a reflecting boundary rather than an absorber? The logic is the same in this case, but one of the boundary conditions is different. If, for instance, the wall at $x = b$ reflects rather than absorbs, we can incorporate this fact in our calculations by noting that at a reflecting surface $dT/dx = 0$. To see the sense of this, consider a particle at the reflecting boundary, b. If the particle takes a step back toward the origin, the time until it is captured by the absorbing wall from this new position is $T(b - \delta)$. If, instead, the particle attempts to take a step away from the origin, it is reflected by the wall and ends up at $b - \delta$, again with a time to capture of $T(b - \delta)$. Because at the reflecting wall the expected time to capture does not depend on whether the particle steps right or left, we conclude that dT/dx must be zero at the wall.

Returning to eq. (6.14) and applying the boundary condition $dT(b)/dx = 0$, we find that in the case of a reflecting wall, $k_1 = -b$. Inserting this value, integrating again, and applying the boundary condition that $T = 0$ at $x = 0$, we find that $k_2 = 0$ and

$$T = \frac{1}{2D}(2bx - x^2), \quad \text{one wall reflecting.} \tag{6.20}$$

This expression is also graphed in figure 6.2. In this case, the time to capture again increases parabolically with distance away from the absorbing wall at the origin, but it reaches a maximum not at $b/2$, but rather at $x = b$, the reflecting wall. This maximum time to capture is $b^2/2D$, four times that found when both walls are absorbing. In fact, for any starting position x, the time to capture is larger if one wall is reflecting than if both walls are absorbing. The overall average time to capture with one wall reflecting is

$$\text{overall } T = \frac{b^2}{3D}, \quad \text{one wall reflecting.} \tag{6.21}$$

Again, time increases as the square of distance because (for a diffusive process) distance traveled increases with the square root of time.

As intriguing as these calculations might be to a mathematician, it may not be obvious what relevance they have to the real world. In fact, the concepts of diffusion to capture can be applied to a wide variety of biological problems, two of which we now explore.

6.2 *Adrift at Sea: Turbulent Mixing of Plankton*

Consider the plight of phytoplankton in a tropical lagoon. In many ways, life is good for these free-floating plants. The sun provides plenty of energy for photosynthesis, and the nitrogenous wastes from the adjacent atoll keep the nutrient concentrations in the water at a healthy level. There is, however, the problem of the clams on the seafloor. Any plankter that comes too close to the bottom is engulfed and eaten. This would not necessarily be a problem if plankton could easily maintain their position in the water column. In this case, a cyanobacterium or diatom that found itself at a comfortable distance above the clams need only hold that position to stay safe.

The ocean is not that accommodating, however. First, there is the problem of sinking. Seawater has a density of about 1025 kg m^{-3}, but phytoplankton cells are typically more dense, about 1055 kg m^{-3}, and as a result they tend to sink. This can be a problem for large cells (those with a diameter of 100 μm or more), which sink at the rate of 3–5 m per day (Mann and Lazier 1991). If the lagoon is 10 meters deep, these cells would be doomed to become clams' chowder within a couple of days. Sinking rate is a function of cell size, however: the smaller the cell, the slower it sinks. For small phytoplankton such as the cyanobacteria (with diameters on the order of 1 μm), sinking rates may be as slow a 3–5 cm per day, and a cell born high in a 10-meter-deep water column could expect to spend a blissful life of 200–333 days before meeting its inevitable downfall.

Unfortunately for phytoplankton, sinking is not the only means of transport. As wind blows over the lagoon, its interaction with the water causes the water to eddy and swirl. These random, turbulent motions of the water can act to mix phytoplankton through the water column in very much the same fashion that the random, thermal motions of molecules cause them to diffuse. Indeed, turbulent mixing (also known as *eddy diffusion*) has long been modeled as a macroscopic analogue of molecular diffusion. Where turbulent motions of a fluid result in random mixing, the flux of materials can be described by the turbulence equivalent of Fick's first equation:

$$J_x = -K_x \frac{dC}{dx}. \tag{6.22}$$

Here J_x is again the flux, and dC/dx is the concentration gradient along the x-axis. In this case, D (the diffusion coefficient) is replaced by K_x, the turbulent (or eddy) diffusivity. Note the subscript x on the turbulent diffusivity. Unlike the molecular diffusion coefficient, which under most circumstances is the same regardless of the axis along which a molecule moves, turbulent diffusivities are often quite different in different directions. The subscript here tells us that this particular diffusivity applies only along the x-axis. We now take advantage of the analogy between thermal agitation and turbulent motion to ask a pertinent question: How long can a phytoplankter expect to survive before the random motion of the water delivers it to a clam?

The answer to this question is easily forthcoming once we realize that we are dealing with a case of simple diffusion to capture. First, although phytoplankton are free to move about in three dimensions, it is only their vertical motion that concerns us here. Thus, the complex motion of the organisms can be dealt with as a random walk along an axis extending straight up from the seafloor. Second, in this case, the term "capture" can be taken quite literally. Any plankter that strikes the seafloor is assumed to be consumed, and the seafloor thus constitutes an absorbing boundary. What about the sea surface? If we ignore such complications as surface tension, we are probably safe in assuming that the sea surface acts as a reflecting boundary—any cell that by chance comes in contact with the surface is deflected back down into the water. The situation shown in figure 6.3 is therefore the same as that shown in figure 6.1, in this case with the axis's origin located at the seabed and the reflecting boundary located a distance b away at the water's surface.

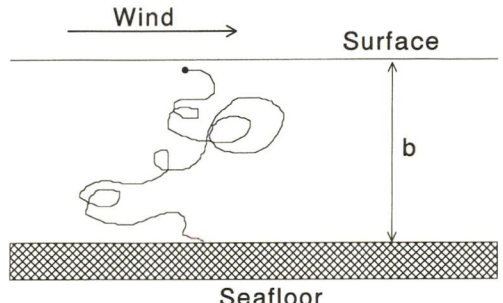

FIG. 6.3 Courtesy of turbulence, a negatively buoyant particle takes a circuitous path as it sinks through a lagoon.

All we need now is the appropriate diffusion coefficient. We will take a simple approach and extract a value for the vertical diffusivity from the literature. Under a wide variety of circumstances, the vertical eddy diffusivity in the ocean's mixed layer is approximately $1 \times 10^{-4} \, \mathrm{m^2 \, s^{-1}}$ (Mann and Lazier 1991). Plugging this value into eq. (6.21) and again assuming a depth of 10 meters, we find that the time to capture for a plankter (averaged over the whole water column) is 3.3×10^5 s, about 3.8 days. Thus, while a plankter can cope with the problem of sinking by being small (resulting in a time to capture from sinking alone of as much as 330 days), this strategy is doomed to failure because turbulent mixing will deliver the plankter to the clams in a much shorter period.

In making this calculation, we have assumed that the vertical diffusivity is the same throughout the water column. In reality this is unlikely to be the case. Eddies created at the water's surface are likely to lose some of their energy to viscosity by the time they have wandered deep into the lagoon, and large eddies that approach the bottom too closely are disrupted by their interaction with the solid substratum. As a result, it is reasonable to expect that K_x will decrease with increasing depth. The value we have calculated here is therefore likely to be an underestimate of the actual time to capture. For our purposes, this rough estimate is sufficient, and we will not delve further into the complex issue of spatially variable eddy diffusivities. The interested reader may wish to consult Csanady (1973) or Mann and Lazier (1991).

Let us now shift gears and explore a different type of random motion.

6.3 *Genetic Drift*

When organisms reproduce sexually, they are in effect performing a set of Bernoulli experiments, though you may not have thought of sex in quite these terms. Consider an example. A diploid organism has two alleles (*a* and *A*) for a particular genetic locus. As chromosomes segregate during meiosis, each gamete receives one or the other allele (but not both). Thus, we can think of the segregation of alleles at this locus as a random experiment consisting of a single trial, again similar to the flipping of a coin.

An interesting question then arises. Consider a population of diploid organisms in which the frequency of *a* and *A* alleles is each 0.5. If these organisms mate at random, what will the gene frequencies be in the next generation? Well, on average half the eggs and sperm will have *A* alleles and the other half will have *a*, and one might conclude that when, in the process of fertilization, we draw a random sample from this gene pool, the frequency of alleles in the resulting zygotes will stay at 0.5. As we have seen, however, we cannot assume that any one random sample will conform to the average, especially if the sample is small. There is always the chance of a "sampling error"—a chance, for instance, that more (or less) than half the eggs and sperm involved in fertilization have *A* alleles.

Consider, for instance, a sample of four gametes (two zygotes' worth) drawn from a population with equal frequencies of *A* and *a* alleles. If these four gametes are drawn at random, there are three possible combinations of alleles in the resulting zygotes (*AA*, *Aa*, and *aa*) and six ways in which these combinations can appear in the two offspring (table 6.1).

In only two of these six cases (*AA* and *aa* zygotes or *Aa* and *Aa* zygotes) is a gene frequency of 0.5 maintained for the *A* allele, and the chance (calculated in chapter 3 using the binomial distribution) of getting one or the other of these cases is only 37.5%. Or, to put it another way, if our population consists of only two newborn individuals, it is very likely (a 62.5% chance) that the gene frequency in the two offspring will *differ* from that in their parents. This potential shift in gene frequencies due to random sampling of gametes is known as *genetic drift*.

Now, genetic drift can only go so far. If by chance *all* the alleles selected during fertilization are of one type (either *A* or *a*), all subsequent sampling error is precluded. For instance, if a generation is produced in which only *A* alleles are present, no matter how this generation's gametes are sampled, the resulting offspring can only have A alleles. Thus, barring mutation, once a given allele has by chance become "fixed" in the population, alternate alleles are permanently lost. It is at this point that our knowledge of diffusion comes into play. By analogy, we can think of a population as performing a random walk along the axis of gene frequencies, shifting its position from one generation to the next. In this analogy, fixation of an allele corresponds to capture by an absorbing wall at a gene frequency of either 0 or 1. We can now put our exploration of "diffusion to capture" to work by calculating how long genetic drift can continue before we expect an allele to "run into the wall" and become fixed or lost.

6.3.1 A Genetic Diffusion Coefficient

The process is surprisingly simple. First, we note that the frequency of an allele is, in essence, an average:

$$\text{frequency} = \frac{1}{n} \sum_{i=1}^{n} A_i, \tag{6.23}$$

where n is the total number of alleles in the population. Here a value of 1 is recorded for A_i if allele A is chosen and 0 if allele a is chosen. From this perspective we can see that by following gene frequencies through time, we are in fact following the variation in an average.

When we start our experiment, the frequency of allele A is p and the frequency of allele a is q. If we then perform a Bernoulli experiment by choosing a gamete at random, the probability is p that it has allele A and q that it doesn't. As we calculated in chapter 4, the variance associated with this type of Bernoulli experiment is pq.

Now an individual gamete is "chosen" when it fuses with a gamete of the opposite gender and forms a zygote. As a result, for the N diploid organisms

TABLE 6.1 The Probability Distribution
for the Alleles in Two Zygotes

Alleles		Probability
AA	AA	0.0625
AA	Aa	0.25
AA	aa $\Big\rangle$	0.375
Aa	Aa	
Aa	aa	0.25
aa	aa	0.0625

Note: Events are defined by the gene frequency, the fraction of A alleles.

of the new generation, $2N$ gametes are sampled. How much is the frequency of A in this new generation likely to differ from that of the parent generation? To put this another way, how much is the "average" of our sample (the frequency of A in the progeny) likely to differ from the average of the parent generation? Recalling our discussion in chapter 4, we can conclude that the average deviation of gene frequency is the same as the standard error of the frequency of allele A in our newly formed zygotes:

$$\text{standard error of the frequency} = \sqrt{\frac{\text{variance of the population}}{\text{number of samples}}}$$

$$= \sqrt{\frac{pq}{2N}}. \tag{6.24}$$

Now, the standard error is, by definition, the standard deviation of the mean, and the square of a standard deviation is a variance. Thus,

$$\frac{pq}{2N} = \sigma_A^2, \tag{6.25}$$

where σ_A^2 is the variance of the frequency of allele A. But this equation is actually a bit misleading. This is the variance accrued in the course of a generation. The notion that this change in frequency is associated with the passing of time can be incorporated by explicitly noting that

$$\frac{pq}{2N} = \frac{d\sigma_A^2}{dt}, \tag{6.26}$$

where time is measured in generations.

Now for the fun part. In chapter 5 we defined the diffusion coefficient to be half the rate of change of the variance of whatever variable we were following. Here we have the rate of change in variance of the frequency of allele A. If we simply divide it by two, we have (by definition; see eq. 5.23) the "diffusivity" of allele A along the axis of allele frequency:

$$D_g = \frac{1}{2}\left(\frac{pq}{2N}\right), \tag{6.27}$$

with the continued proviso that time is measured in generations. Noting that $q = 1 - p$,

$$D_g = \frac{p - p^2}{4N}. \tag{6.28}$$

This conclusion may seem surprising. In contrast to the molecular diffusion coefficient, the "diffusion coefficient" for gene frequencies is not a constant. Instead, D_g is maximal when $p = 0.5$, and tends toward 0 as frequency approaches either 0 or 1. Upon reflection, however, this variability in D_g makes sense. Consider, for instance, an extreme case in which only one a allele is left in the parent population (that is, p is very near 1). In this case, we can make an "error" in selecting alleles for the next generation if the sole a allele is not included. The probability of making this error is indeed high, but if we do, we have only shifted the gene frequency a tiny amount. In other words, for p near 1, the genetic drift from one generation to the next should be slight. In contrast, when A and a alleles are present in equal numbers, there are many possible ways to make an "error" in sampling, and some of them lead to substantial shifts in gene frequency. Thus, when p is near 0.5, drift can be relatively fast.

6.3.2 DRIFT AND FIXATION

Having arrived at an expression for the diffusivity of gene frequencies, we abruptly find ourselves in a position to estimate an allele's expected time to fixation due to genetic drift. In eq. (6.18) we calculated that the time to capture for a particle confined between two absorbing walls is

$$T(x) = \frac{1}{2D}(bx - x^2).$$

In our genetic analogy, the constant b corresponds to a gene frequency of 1. Thus,

$$T(p) = \frac{1}{2D_g}(p - p^2). \tag{6.29}$$

Inserting our value of D_g (eq. 6.27) into this equation, we arrive at the conclusion that

$$T(p) = \frac{2N}{p - p^2}(p - p^2)$$
$$= 2N. \tag{6.30}$$

In other words, the expected number of generations required to fix an allele by genetic drift is simply equal to twice the number of individuals in the population.[2] The larger the population, the slower the drift.

The conclusion that small populations can rapidly drift to fixation is of considerable practical importance. Consider, for example, a species (such as the

---

[2] To be precise, N is the number of surviving offspring per generation. But in traditional treatments of genetic drift, it is generally assumed that (1) individuals reproduce and then die, and (2) the total number of organisms in a population is constant from one generation to the next. Under these conditions, N is equal to the population size.

cheetah or the sea otter) that has been hunted to near extinction in the wild. There are cheetahs in zoos, and they can be coaxed to breed, but due to the small number of individuals in the zoo population it is likely that many alleles in these animals will drift to fixation over the next few generations. The alternative alleles (those not fixed) are lost for all time. The practical importance of genetic drift is likely to increase as humans reduce the population size of many species.

Before we get carried away examining the effects of genetic drift in small populations, it behooves us to examine eq. (6.30) in greater depth. Doesn't it seem surprising that the time to fixation is always the same, regardless of the initial gene frequency? Indeed, this should be worrisome. It implies, for instance, that because the time to capture is the same from $p = 0.5$ as it is from $p = 0.3$, that it either takes no time to drift from $p = 0.5$ to $p = 0.3$ (which, given our assumptions, cannot be true), or that in drifting from 0.5 to 0.3, some "momentum" is attained that affects subsequent drift. In fact, because of the way we have defined the rules of a random walk we have (incorrectly) imposed a kind of "momentum" to the changing frequency of alleles. This effect is best explained using an example.

In chapter 5, we defined a diffusion coefficient for a one-dimensional random walk in terms of a step length, δ, and the interval between steps, τ.

$$D = \frac{\delta^2}{2\tau}. \tag{6.31}$$

We can rearrange this equation to express the step length as a function of D and τ:

$$\delta = \sqrt{2D\tau}, \tag{6.32}$$

an expression that can be translated into values appropriate for the drift of gene frequencies. Given a stepping interval of one generation and D_g as defined by eq. (6.28),

$$\delta = \sqrt{\frac{p - p^2}{2N}}. \tag{6.33}$$

In other words, the step length varies as a function of allele frequency, p, and population size, N. δ is largest when $p = 0.5$ and decreases for allele frequencies that are either larger or smaller.

To see the practical result of this variable step size, let's follow the path of a genetic "particle" that is constrained to take alternating steps to the left and the right along the axis of allele frequencies. Let $N = 4$. If the particle starts at $p = 0.5$, the step length (calculated from eq. 6.33) is 0.1768, and with its initial

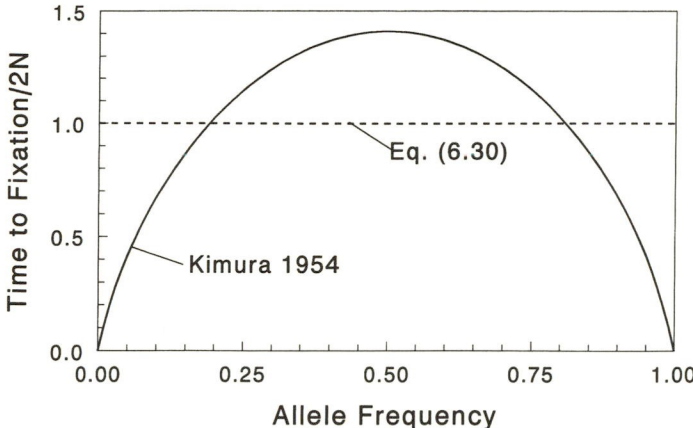

FIG. 6.4 In an accurate model of genetic drift (Kimura 1954), the time required for an allele to drift to fixation depends on the starting allelic frequency; it is highest when the initial frequency is 0.5. The estimate calculated here (a time of $2N$ generations, independent of starting frequency) is equal to the time to fixation averaged over all starting frequencies.

step to the left it arrives at $p = 0.3232$. The particle then takes a step back to the right, but at its current position, the step length is only 0.1654, so this step takes it only to $p = 0.4886$. In other words, even though the particle has taken an equal number of steps to the left and the right, it does not return to its starting position. The next step to the left takes the particle to $p = 0.3119$, and the subsequent step to the right returns it only to $p = 0.4757$. As the particle steps back and forth, it gradually drifts toward $p = 0$. This is the "momentum" referred to above. If we had started the particle at $p = 0.5$ but given it an initial step to the right, the particle would drift toward $p = 1$. Thus, the net result of a diffusivity that varies along the axis of motion is a tendency for particles to drift toward the area(s) of the axis where diffusivity is lowest. It is this insidious drift that leads to the strange result of eq. (6.30).

How could we have led you so astray? It turns out that our rough-and-ready result is not all that far off the truth. Kimura (1954) calculated the actual rate at which alleles are lost by drift with the results shown in figure 6.4.[3] These results (as well as being correct) are also more intuitive. The time to fixation varies

[3] Unfortunately, the equation describing these results is expressed in terms of an infinite series of Legendre polynomials. It isn't difficult to teach a computer to make the appropriate calculations, but the equation is a bit too forbidding to include here. If you feel adventuresome, consult Kimura (1954) for the equation (his eq. 8) and Abramowitz and Stegun (1965) for the recursion formula for calculating Legendre polynomials.

from a maximum of approximately $2.8N$ generations for a starting frequency of 0.5 to a minimum of 0 for a starting frequency of 0 or 1. Averaged across all frequencies, however, the expected time to fixation is $2N$ generations, the same answer we arrived at with our naive approach.

For a discussion of the effects of spatially variable diffusivity, consult Schnitzer et al. (1990) or McNair et al. (1997).

6.4 *Genetic Drift and Irreproducible Pigs*

Our discussion of genetic drift provides us with the background for a brief side trip to the interface between random walks and the algebra of probability, and to a consideration of the troubling subject of irreproducible results. We begin with a return to the type of simple thought experiment we employed in chapter 2.

Consider the following experiment (referred to in the literature of statistics as "Polya's urn scheme"). Two balls, one white and the other black, are placed in a large urn. The balls are identical except for their color, and can thus be retrieved from the urn at random. If the ball you pull out is white, it is returned to the urn, and another white ball is added. If the ball you retrieve is black, it too is returned to the urn, but in this case a black ball is added. This process is repeated; at each step a new ball is added to the urn, its color depending on that of the ball retrieved. What happens to the ratio of white to black balls as the number of repetitions mounts?

A few repetitions of this experiment (two hundred balls' worth or so) are enough to convince one that, after some initial variability, the ratio of white to black balls quickly approaches a particular value and thereafter remains virtually constant. In other words, the ratio quickly becomes predictable. The unusual aspect of this experiment is that each time the experiment is carried out, a different ratio is obtained! Results that are accurately reproducible within an experiment (the ratio of white to black balls quickly becomes constant) are irreproducible among experiments (different constant values are obtained in trials of the same experiment).

A practical example of this type of phenomenon was presented by Cohen (1976), who considered the interaction of genetic drift with the breeding of pigs. Despite their "good looks and intelligence," pigs are raised primarily for their meat. As a result, most members of a given generation are slaughtered; only a few are spared to produce the next generation. As we have just calculated, genetic drift may occur rapidly in such a small breeding population.

Now, consider one of the possible consequences of this drift. A characteristic such as weight at adulthood is commonly controlled by the additive effect of several genes. For example, we can imagine a situation in which there are five genes contributing to the overall marketable mass of a pig, each gene having two alleles. The presence of one of each of these alleles (which we shall denote with a capital letter) leads to an increase of one pound above the average; the presence of the alternate allele (shown in lower case) results in a decrease of one pound. So, for instance, a pig with the homozygous genotype *AABBC-CDDEE* will weigh 10 pounds more than an average-weight pig with the uniformly heterozygous genotype *AaBbCcDdEe*, and 20 pounds more than a pig with the genotype *aabbccddee*.

Suppose that Farmer Bob starts his pig population with two pigs, each heterozygous at each of these five loci. Each spring piglets are produced in porcine profusion, but only two (chosen at random) are spared for breeding purposes. What happens to the weight of Bob's porkers as the generations go by?

Well, we have shown that it is highly likely, given the small size of the population of Bob's breeding pigs, that random drift will result in the fixation of one or the other allele at each locus through the course of just a few generations. After this time, all Bob's pigs are homozygous. They will all weigh the same within a generation, and this weight will be constant across generations. This much we can predict. What we *cannot* predict is what this constant weight will be. It is just as likely that the first locus will drift to fixation at a value of *AA* as it is for a value of *aa*, and the same is true for the other four loci. As a result, it is just as likely that Farmer Bob will end up with a true-breeding population of runts as it is that he will acquire a true-breeding set of prize-winning heavyweights. Thus, we have a rough real-world analog to Polya's urn scheme. Had Farmer Alice started with a pair of pigs identical to that of Farmer Bob, she too would quickly have a true-breeding population of pigs, but the weight of Alice's pigs is likely to be different from the weight of Bob's. Within each farm, the weight of pigs becomes predictable; across farms, the results of this repeated experiment are irreproducible.

Other examples of this phenomenon are easy to imagine. For instance, many benthic animals have planktonic larvae, and these larvae will settle only near members of their same species. Consider the plight, then, of the poor field ecologist who is trying to study the competitive abilities of two barnacle species, B_1 and B_2, in particular their ability to compete for space during settlement. Bare settlement plates are introduced into the environment, each plate bearing a single adult of each species—a reasonable example of a controlled, repeated experiment. But if the first larva that happens to contact a plate is from species B_1, it settles, thereby doubling the chance that the next S_1 larva will contact a member of its own species and likewise settle. This first, random encounter thus biases the rest of the experiment, and this particular plate may end up with a high ratio of B_1 to B_2 individuals. If, however, the first larva that happened by was of species B_2, a different outcome could result. Thus, what would appear to be an unbiased competition experiment in which similar results might be expected among repetitions may turn out to have a high degree of interplate variability. Furthermore, this variability has nothing to do with the competitive abilities of barnacles; it is simply a matter of chance.

For further examples of Polya's urn phenomenon and related phenomena, the interested reader is urged to consult Cohen (1976), May (1976), or Dunthorn (1996).

6.5 *The Biology of Elastic Materials*

It is now time to step beyond the statistics of random motion in one dimension to explore the consequences of three-dimensional random walks. We will, however, approach the subject from what may seem like a strange direction by first exploring the biology of elastic materials.

6.5.1 ELASTICITY DEFINED

Consider a rubber band. If you hang a weight from one end of the band, the band extends, and this new length is maintained as long as the weight is suspended. The more weight you apply, the more the band stretches. If all weight is removed, the band immediately returns to its unstretched length. These are the properties of an *elastic* material, and they are summarized in figure 6.5. The deformation of a piece of the material is proportional to the force applied. In the simple case shown here, the ratio of force to deformation is constant and is known as the material's *stiffness*. Expressed as an equation (known as Hooke's Law),[4] the relationship between force and deformation in an elastic material is

$$\text{force, } F = \mathcal{K}x, \tag{6.34}$$

where $\mathcal{K}$ is the stiffness of a particular bit of the material (expressed in newtons per meter), and x is the sample's deformation. Here we assume that $\mathcal{K}$ is constant, but in many biological materials $\mathcal{K}$ is a function of both the absolute amount of deformation and the deformation rate (Wainwright et al. 1976).

---

[4] Named for Robert Hooke (1635–1703). Through a series of experiments with springs, Hooke was able to document the linear relationship between force and deformation that characterizes many materials (Timoshenko 1953). As curator of experiments for the Royal Society of London, Hooke was intimately involved in the scientific renaissance of the seventeenth century, but not always for the best. He rudely criticized Newton's work on optics and misused his office with the Royal Society, causing Bell (1937) to refer to him as "the pestiferous Hooke."

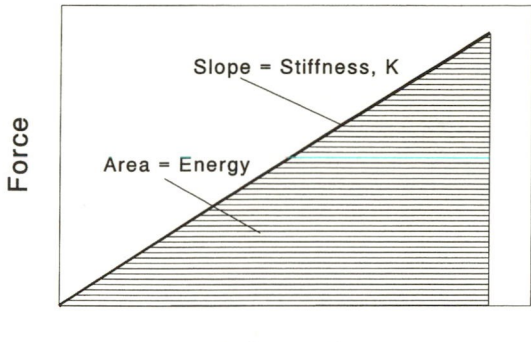

FIG. 6.5 A force-deformation curve for an elastic solid. The stiffness of the material depends on how fast force rises with increased deformation. The elastic potential energy stored in an elastic material is equal to the area under the force-deformation curve.

Now, mechanical energy is defined as the product of force and the distance through which the force moves its point of application (in this case, the deformation of a material). If force varies with deformation, this is best expressed as

$$\text{energy} \equiv \int F(x)\,dx. \tag{6.35}$$

Thus, the energy expended in deforming an elastic object to a deformation x is (by the definition of the integral) the area under the force-deformation curve between 0 and x, as shown in figure 6.5.

If a material is truly elastic, all of the energy expended in deforming the material is stored as potential energy and can be used to do work. For example, in grammar school you probably discovered that the energy stored in a stretched rubber band could be used to propel the band at a satisfying speed toward the neck of the student in front of you. The elastic potential energy stored in stretched metal springs powers wind-up clocks and assists you when opening garage doors. It is this ability to do work that makes elastic potential energy so useful to biology.

6.5.2 BIOLOGICAL RUBBERS

As a pertinent example, consider flying insects. More than half the animal species on earth are insects, and most of these fly. To propel themselves off the ground, flying insects beat their wings, thereby pushing air downward. This downward jet of air provides the lift that supports the insect's weight against the pull of gravity. In the course of beating its wings, however, an insect faces a common problem in the energetics of locomotion. Some of the energy expended in beating the wings goes into the downward acceleration of air and, by providing lift, does useful work. But another portion of the energy expended in a wing beat is wasted. For example, as a wing is flapped up and down, a force

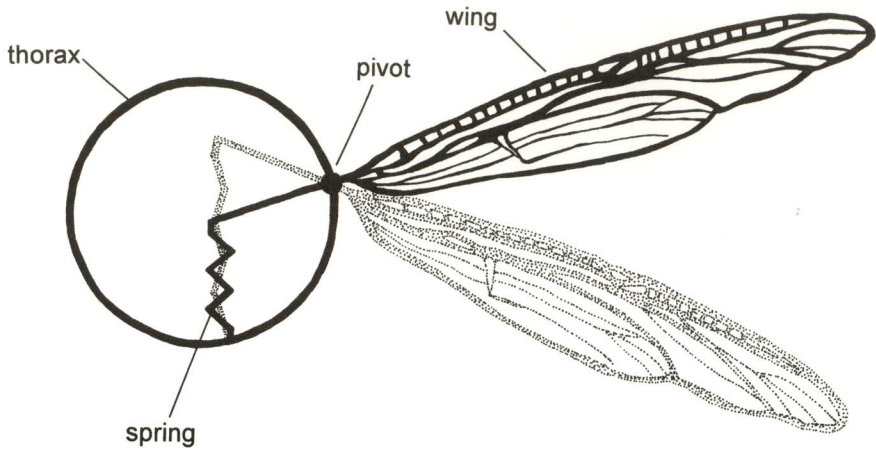

FIG. 6.6 A schematic diagram showing how a spring can act to assist the flapping of an insect's wing.

is required first to get the wing's mass up to speed and then to slow the mass down so that its direction of motion can be reversed. It would be possible for all of this force to be supplied by muscles, but the production of force by a muscle requires the expenditure of energy in the hydrolysis of ATP. Thus, the flapping of wings by muscles alone would be energetically costly even if the wings were producing no lift.

There is, however, a way around the expenditure of muscular energy to accelerate and decelerate wings. Consider the situation shown schematically in figure 6.6. Here, the wing is attached to a spring. As the wing moves down, the spring is stretched, and as a result a force is applied to the wing, causing it to decelerate. In this sense, the action of the spring is similar to the action of a muscle. Unlike in a muscle, however, elastic potential energy is stored in the spring's material. Thus, when the wing reaches the bottom of its beat, the energy stored in the spring can be used to accelerate the wing upward. As the wing moves upward, it eventually begins to compress the spring. The force of compression decelerates the wing in its upward path, and the stored energy of the compressed spring can serve to accelerate the wing downward. In this fashion, once the wing's motion has been initiated, the beating of the wing can be maintained without the need to expend energy accelerating and decelerating the wing's mass. Admittedly, in a real, functioning wing muscular energy would have to be expended to overcome the friction between the wing, its pivot, and the surrounding air, and to accelerate enough air downward to provide lift. But the energy saved by allowing elastic potential energy to accelerate the wing's mass is considerable (see Alexander 1992 for a review).

As you might expect, in the course of evolution some insects have taken advantage of the energetic savings afforded by the storage of elastic potential energy, and have mounted their wings on springs. To be more precise, the springs in wings take the form (at least in part) of tendons and hinge pads formed from a protein rubber called *resilin*. The name is intended to convey the fact that this particular rubber is highly efficient in storing elastic energy, and is therefore very resilient.

Another biological example of potential energy storage is found in the bivalves—the clams, cockles, mussels, and scallops of the world. These mollusks avoid predation and desiccation by enveloping their bodies with two hard, calcareous shells. When threatened, they "clam up," pulling their shells together and leaving no living material exposed to the outside world. The force required to pull the shells together (to adduct the shells, in the terminology of such things) is provided by large adductor muscles (fig. 6.7). These muscles form one of the tastier portions of a bivalve, and in a scallop are the only part that is commonly eaten.

To this point, the mechanical strategy of the bivalve seems straightforward and effective. There is only one problem. Having pulled the shells together, how can the clam, mussel, or scallop get them back apart? Muscles are no help here; they can pull but not push.[5] The answer, as you might suspect, is found in the presence of an elastic material in the shell's hinge (fig. 6.7). When the shell is adducted, this rubbery material is either stretched or compressed, and elastic potential energy is stored. When the adductor muscles are relaxed, this stored energy is used to push the shells apart in a process known as *abduction*. The protein rubber used by bivalves to open their shells is appropriately called *abductin*, and in older invertebrate texts, the pad of abductin found inside the shell (as in scallops) is called a *resilium*, a wonderfully descriptive name.

In vertebrates, rubbery materials are used to store elastic energy in a variety of situations. For example, the heart of a vertebrate is a pulsatile pump—every time the ventricles contract, a bolus of blood is ejected into the arteries. A consideration of the energetics of flow in pipes, however, shows that it costs less to move blood through the body's plumbing if the flow is steady rather than pulsatile (Alexander 1983). The conversion of the intrinsically pulsatile flow produced by the heart to the steady flow that is advantageous in the capillaries is brought about by the elastic properties of the arteries, particularly the rubbery behavior of the large arteries near the heart. As the heart contracts, blood

[5] Under certain circumstances, muscles can be used to provide a pushing force (for a discussion, see Smith and Kier 1989). Bivalves have not taken advantage of this possibility, however.

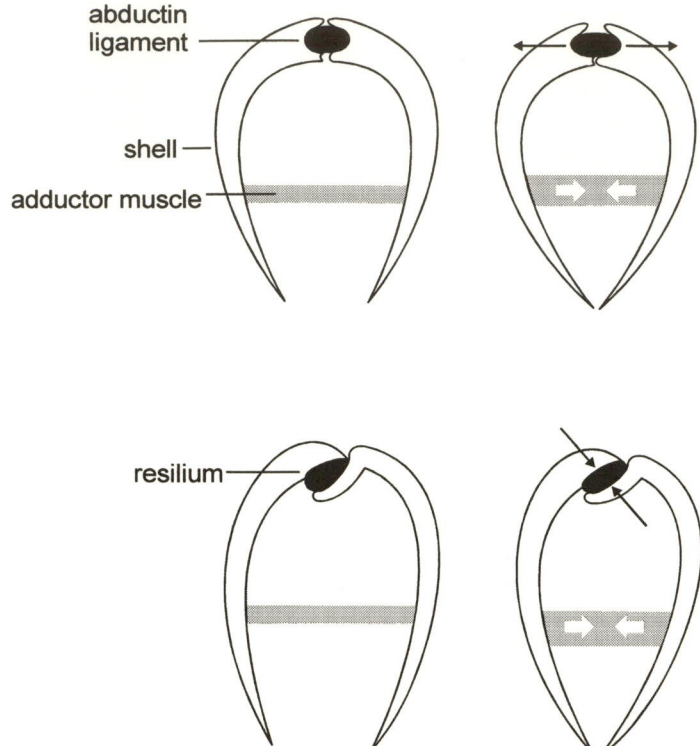

FIG. 6.7 Elastic potential energy can be stored in the hinge ligament of a bivalve. This ligament can either be external to the shell (as in the upper shell) or internal (as in the lower shell). The external ligament is stretched, and the internal resilium is compressed as the adductor muscles contract and the organism "clams up."

is forced into the largest of these, the aorta. Some of this blood immediately flows on into the peripheral vascular system, but the remainder is stored locally as the walls of the aorta are stretched. Because the artery walls contain a protein rubber (in this case, *elastin*), the energy used to stretch the walls is stored. When the heart stops pumping, the now swollen aorta contracts, in effect acting as a second pump that keeps the blood flowing while the heart reexpands. The net result is flow in the peripheral arteries and capillaries that is much more steady than it would otherwise be, and a considerable energy saving to the organism.

A similar scheme is found in octopuses and squids. These cephalopod mollusks are the only invertebrates with a high-pressure, closed-circuit vascular system analogous to that of the vertebrates. As such, these organisms face the same mechanical problems in pumping blood through their bodies as do vertebrates, and their evolution has converged on a similar solution. In this case, the elastic protein in the vessel walls is given a less catchy name—*octopus arterial elastomer* (Shadwick and Gosline 1983).

ligamentum nuchae

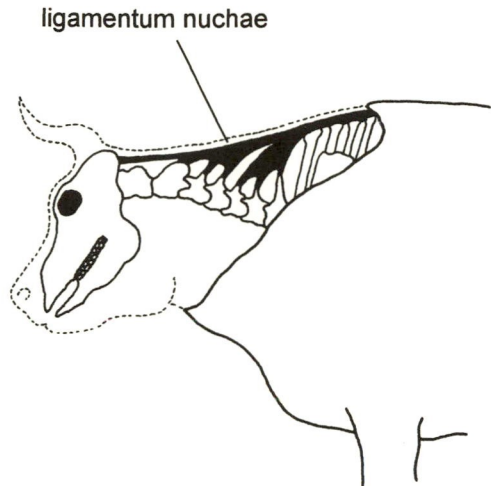

FIG. 6.8 The ligamentum nuchae (a large band of elastin) helps to support the neck and head of cows (shown here), horses and other large mammals.

Elastin is also used to support the heads of large ungulates. For example, the head of a cow is a sizable mass perched on the end of a horizontal vertebral column. If the head were held up by muscles, considerable energy would need to be expended to maintain a "heads up" posture. Instead, much of the mass of the head is held up by a large elastic ligament (the *ligamentum nuchae*) that runs from the dorsal spines of the thoracic vertebrae to the cervical vertebrae and the base of the skull (fig. 6.8).

6.5.3 THE LIMITS TO ENERGY STORAGE

Given that the storage of elastic potential energy has a wide variety of advantages in biology, why are only a few structural materials (primarily the protein rubbers) used for this purpose? What is special about the energy storage abilities of rubbery materials? To answer this question, we need to examine the relationship among stiffness, deformability, and elastic energy. The maximum elastic energy that can be stored by an object is equal to the energy required to deform the object to its maximum extension, x_{max}:

$$\text{energy} = \int_0^{x_{max}} F(x)\,dx$$

$$= \int_0^{x_{max}} \mathcal{K}x\,dx$$

$$= \frac{1}{2}\mathcal{K}x_{max}^2. \tag{6.36}$$

The amount of energy that a given piece of material can store therefore depends on both the material's stiffness and on how far it can extend. However, because the extension is raised to the second power, the overall storage capacity of an object is more sensitive to its ability to extend than to its stiffness. For example, elastic energy can be stored in bones, which are elastic and quite stiff but can extend by only a fraction of a percent of their unstretched length before they break (fig. 6.9). In comparison, a piece of protein rubber of similar size is much less stiff, but can easily extend to 2–3 times its unstretched length. As a result, the area under the force-extension curve (a measure of the stored elastic energy) is larger for the rubber than for the bone. Thus, as regards their use as energy storage devices, biological rubbers are notable for two features: they are indeed elastic, and they have large maximal extensions.

We are now presented with an intriguing question. How can we explain the ability of protein rubbers to stretch to several times their length and still return elastically to their initial dimensions? The answer is by no means obvious, but given the importance of rubbery materials in biology, it behooves us to solve the riddle. It is at this point that biology crosses paths with the statistics of a three-dimensional random walk.

A brief warning is in order. The analysis that follows is a bit tortuous, involving three disparate lines of thought. Please bear with us. In the end, all the

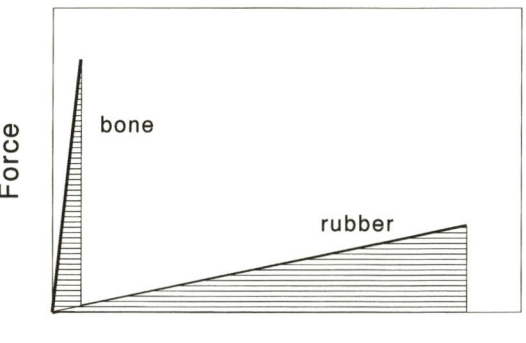

FIG. 6.9 The elastic energy stored in a material is the area under the force-deformation curve. If the material is strong and stiff but not very extensible (such as bone), relatively little energy is stored. A large amount of energy can be stored if the material is very deformable (as in rubber), even if it is not stiff or strong.

pieces fit together, and we feel that the elegance of the answer is worth the work. The logic here follows that of Aklonis et al. (1972).

6.6 *Random Walks in Three Dimensions*

The basic ideas of a random walk are easily transferred from one dimension to three. As before, particles are released at the origin of our coordinate system, and they still take steps of a fixed length δ at intervals of τ seconds. The direction of these steps is still random. "Random," in this case, involves an entirely different degree of complexity, however. A random step on a single axis involves the choice between only two locations: a particle either moves one step to the right or one step to the left. Although it is impossible to tell beforehand to which of these two locations the particle will move, the results are relatively simple to envision. If, however, a particle is free to move in three dimensions, the taking of a single random step involves the choice among infinitely many locations. At the end of the step the particle can be located anywhere on the surface of a sphere one step length away from its previous position.

To specify the particle's position through time we need to know the x, y, and z coordinates of the particle after each step. It will prove useful to keep track of these coordinates not as separate entities, but rather as they are combined to describe a line, $\vec{r}$, which extends from the origin ($x = 0$, $y = 0$, $z = 0$) to the point x, y, z where the particle lies (fig. 6.10). This line (technically, a vector) has a length and a direction, and therefore contains all the information regarding the location of the particle. For the moment, however, we do not worry about the direction of $\vec{r}$, we consider only its length:

$$\text{length of } \vec{r} = r = \sqrt{x^2 + y^2 + z^2}. \tag{6.37}$$

When we use r (instead of $\vec{r}$), we are thus limiting ourselves to information regarding how far the particle has wandered from its starting point, forgoing information about the specific direction in which it has moved.

In much the same fashion as we calculated the average distance moved by a particle in a one-dimensional walk, it is possible to calculate the average distance moved in three dimensions. The calculation is somewhat complex, however, and we do not present it here because the result has such a familiar "feel" to it.[6] The mean square distance traveled by a particle in a three-dimensional random walk is

$$\overline{R^2} = n\delta^2, \tag{6.38}$$

---

[6] The interested reader should consult Aklonis et al. (1972) for a readable derivation.

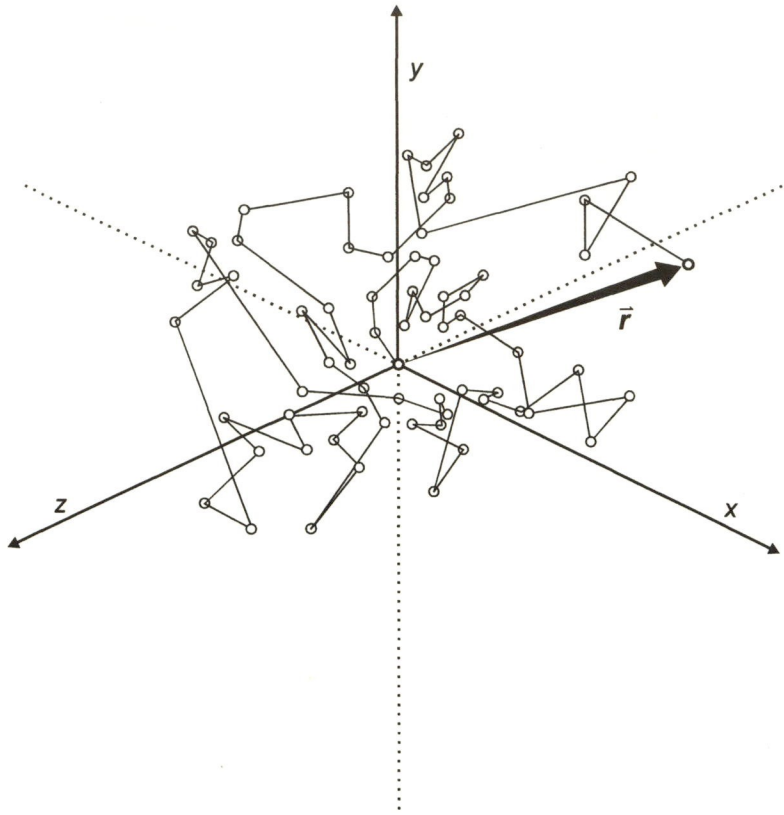

FIG. 6.10 A random walk in three dimensions. The vector r connects the origin (the start of the random walk) to the particle's location after a number of steps.

where n is the number of steps taken and δ is again the step length. This is quite similar to the conclusion we reached for motion in one dimension, where $\overline{X^2} = n\delta^2$.

For each step of a fixed length, the choice of a direction in which to move means that motion in the x direction is tied to that in the y and z directions. But the direction of motion is random from one step to the next, so, taken across many steps, motions along the three coordinate axes are independent. This statistical independence is important in the present context because it allows us to apportion the overall mean squared distance $\overline{R^2}$ among the motions along the different axes. Recall that in chapter 3 we showed that if variables are independent, the variance of the sum of variables is equal to the sum of the variances of the individual variables. Here the mean squared distance $\overline{R^2}$ is analogous to a variance, and we can rightfully assert that the overall mean square distance is equal to the sum of the mean square distances traveled along each axis:

$$R^2 = \overline{X^2} + \overline{Y^2} + \overline{Z^2}. \tag{6.39}$$

If the choice of direction for each step is truly random, there is no reason to think that motion along any one axis is different from that along another, which leads to the conclusion that

$$\overline{X^2} = \overline{Y^2} = \overline{Z^2} = \frac{\overline{R^2}}{3}. \tag{6.40}$$

Now, we know from our consideration of one-dimensional random walks that in the case of a large number of steps we can describe the probability of a particle being located at a given point on the x-axis using the normal distribution. For the present situation, in which the mean displacement along an axis is 0 (that is, $\mu_X = 0$), the normal probability density distribution can be expressed as

$$g(X) = \frac{1}{\sqrt{\overline{X^2}}\sqrt{2\pi}} \exp\left(-\frac{x^2}{2\overline{X^2}}\right), \tag{6.41}$$

where $\overline{X^2}$ is the variance in particle location and $g(X)$ is again the probability density function. By substituting $\overline{R^2}/3$ for $\overline{X^2}$ we can describe the probability density of a particle's location on the x-axis in terms of r, the particle's overall three-dimensional distance from the origin:

$$g(X) = \frac{\sqrt{3}}{\sqrt{\overline{R^2}}\sqrt{2\pi}} \exp\left(-\frac{3x^2}{2\overline{R^2}}\right). \tag{6.42}$$

Analogous equations can be used to describe the probability density of particle locations along the y- and z-axes.

Recall that the probability density of x is the probability *per distance* along the x-axis. By analogy, we can think of the probability density for motion in three dimensions as the probability *per volume*. In other words, by using the appropriate function, we can describe the probability that a particle will be found in a particular volume in space. For example, we might want to know the probability that a particle will be found in the small volume $dxdydz$ located at the end of vector $\vec{r}$. In fact, knowing the probability density function for each axis (which we have just calculated) and the fact that motion is independent along the three axes, we can calculate the necessary three-dimensional probability density function with surprising ease. It is simply the product of the three individual probability density functions:

$$g(X, Y, Z) = g(X)g(Y)g(Z). \tag{6.43}$$

This multiplication is easily carried out by first taking the logarithm of each side of the equation:

$$\ln[g(X, Y, Z)] = \ln[g(X)] + \ln[g(Y)] + \ln[g(Z)]$$

$$= \left[3\ln\left(\frac{\sqrt{3}}{\sqrt{\overline{R^2}}\sqrt{2\pi}}\right)\right] + \left(\frac{-3x^2}{2\overline{R^2}} + \frac{-3y^2}{2\overline{R^2}} + \frac{-3z^2}{2\overline{R^2}}\right)$$

$$= \left[3\ln\left(\frac{\sqrt{3}}{\sqrt{\overline{R^2}}\sqrt{2\pi}}\right)\right] + \left(\frac{-3(x^2 + y^2 + z^2)}{2\overline{R^2}}\right)$$

$$= \left[3\ln\left(\frac{\sqrt{3}}{\sqrt{\overline{R^2}}\sqrt{2\pi}}\right)\right] + \left(\frac{-3r^2}{2\overline{R^2}}\right). \tag{6.44}$$

Taking the antilog of this equation, we see that

$$g(X, Y, Z) = \left(\frac{\sqrt{3}}{\sqrt{\overline{R^2}}\sqrt{2\pi}}\right)^3 \exp\left(-\frac{3r^2}{2\overline{R^2}}\right) = \frac{dP}{dV}. \tag{6.45}$$

This is a spatial probability density function, the probability per volume.

Now, let's consider not the infinitesimal volume $dxdydz$ at the end of $\vec{r}$, but rather all the volume at the surface of a sphere with radius r. In a sense, we are summing all the individual volumes $dxdydz$ that are r away from the origin, thereby removing the effect of direction. At a distance r from the origin, the volume dV contained in a thin spherical shell dr thick is $4\pi r^2 dr$ (the surface area of the sphere times the thickness of the shell). Thus, the probability that a particle is located within the volume that lies between r and $r + dr$ is

$$dP = g(X, Y, Z)\, dV \tag{6.46}$$

$$= g(X, Y, Z) \cdot 4\pi r^2 \, dr \tag{6.47}$$

$$= \left(\frac{\sqrt{3}}{\sqrt{\overline{R^2}}\sqrt{2\pi}}\right)^3 \exp\left(-\frac{3r^2}{2\overline{R^2}}\right) 4\pi r^2 \, dr. \tag{6.48}$$

Dividing both sides of this equation by dr, we arrive at the radial probability density function $g(r)$:

$$g(r) = \left(\frac{\sqrt{3}}{\sqrt{\overline{R^2}}\sqrt{2\pi}}\right)^3 \exp\left(-\frac{3r^2}{2\overline{R^2}}\right) 4\pi r^2. \tag{6.49}$$

In words, this is the probability per volume that a particle can be found at a given distance r from the origin, where the direction of r is not taken into account. This result, expressed here in terms of r, can be restated in terms of

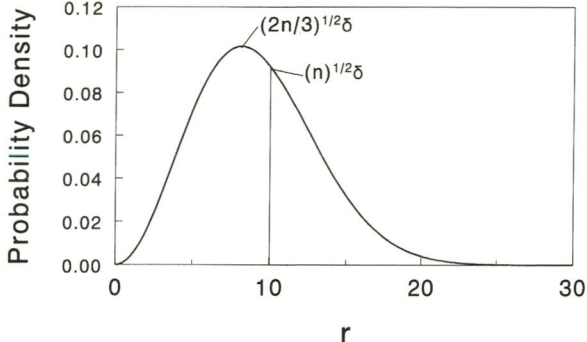

FIG. 6.11 The probability density function of distance traveled for a three-dimensional random walk (eq. 6.50). The average distance $(n^{1/2}\delta)$ is greater than the most probable distance $[(2n/3)^{1/2}\delta]$ due to the long tail on the curve extending to large distances. In the example shown here, $\delta = 1$ and $n = 100$.

the number of steps taken. Earlier we asserted that $\overline{R^2} = n\delta^2$, and substituting this value into eq. (6.49), we see that

$$g(r) = \left(\frac{\sqrt{3}}{\sqrt{n}\delta\sqrt{2\pi}}\right)^3 \exp\left(-\frac{3r^2}{2n\delta^2}\right)4\pi r^2. \qquad (6.50)$$

This expression is graphed in figure 6.11 for a random walk of one hundred steps, each of length 1.

When r is small, the probability density is small. In other words, the probability of finding a particle near the origin is low and at the origin it is zero. This is in contrast to a one-dimensional random walk, where the most probable location of a particle released at the origin is always at the origin. The difference here is that in a three-dimensional random walk, the particle can move in any direction at any step, and therefore is not confined to repeatedly retracing its steps. The likelihood that a particle will *exactly* retrace its steps (or otherwise end up precisely back at the origin) is zero.

The probability density curve rises to a peak at $r = \sqrt{(2n/3)}\delta$, and then declines at larger distance from the origin. There is a long tail to the curve at large r, and as a result the mean distance traveled, $\sqrt{n}\delta$, is greater than the most probable distance, $\sqrt{(2n/3)}\delta$.

6.7 Random Protein Configurations

What can these calculations tell us about biology? We can begin to tie our calculations to nature by considering the structure of a long polymeric molecule such as the proteins from which resilin, abductin, and elastin are formed. Each polymer is a chain of n monomers linked together serially, and each link has a length δ. For proteins, the individual links are amino acids (fig. 6.12), and the size of these molecules sets the size of the link. Ignoring some minor restrictions, the peptide bonds between amino acids are more or less free to rotate, and in the rubbery proteins the side chains of the amino acids have very little tendency to attract, repel, or impede each other. This freely rotating, nonattracting, nonrepelling nature of rubbery proteins imbues these chains with great flexibility, and we can think of the configuration of the protein as a physical depiction of a three-dimensional random walk.[7]

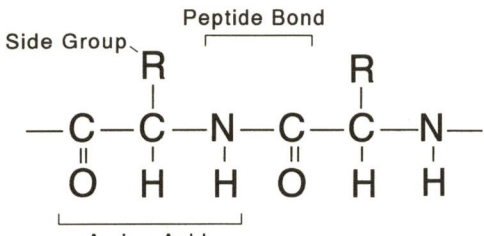

Fig. 6.12 The sequence of atoms in a polypeptide chain (a protein).

Furthermore, we can guess that thermal agitation will continuously rearrange the orientation of each peptide bond, and the precise configuration of the polymeric chain will change through time in an unpredictable fashion. As always when we encounter this type of temporal unpredictability, we can seek some order by examining time-averaged values. In particular, the statistics we have just explored regarding random walks in three dimensions allow us to specify what the average distance is between the ends of a protein molecule.

To see this, imagine that we could tack one end of the protein to the origin of our coordinate system. In this case, the end-to-end distance of the protein is the length of $\vec{r}$, the line that connects the origin to the location of the particle at the end of its random walk of n steps. Because the protein molecule *is* the random walk, the length of $\vec{r}$ must be the same thing as the protein's end-to-end distance. We know that this distance continuously changes as the protein is thermally agitated, but from our calculations we know that its average is $\sqrt{n}\delta$. For example, each amino acid has a length of about 0.5 nm along the protein

---

[7] In contrast to rubbery proteins, the polypeptide chains of most proteins are not nearly as flexible. In a typical protein, sections of the chain are likely to be hydrogen bonded to other sections in alpha helices or beta sheets, disulfide bonds or hydrophobic interactions may hold the chain in a relatively fixed structure, and the bulky side chains of some amino acids constrain the flexibility at individual peptide bonds. In the presence of these hindrances, a typical protein chain cannot be accurately modeled as a random walk.

chain (Lehninger 1970). Thus, if a chain consists of a hundred amino acids, its average end-to-end distance is $\sqrt{100} \times 0.5 = 5$ nm. Furthermore, we know that the most probable end-to-end distance is $\sqrt{(2n/3)}\delta$. Thus, for the same protein of 100 amino acids, the most probable end-to-end distance is 4.1 nm, slightly less than the average distance.

It is interesting to note that when n is large, the average end-to-end distance for a protein molecule is much shorter than the overall length of the chain. For instance, in the hundred-link chain just described, the average end-to-end distance is only 10% of the chain length. This is one indication of the fact that the average configuration of a rubbery protein chain undergoing thermal agitation is fairly compact.

It is also worth noting yet again that it is the random rearrangement of the protein chain that allows us to predict its average dimensions accurately. In fact, it is this *disorder* (and our ability to quantify it) that eventually will allow us to account for the elasticity of rubber.

6.8 *A Segue to Thermodynamics*

Let us now shift gears again.

We are interested in the force F required to maintain a given end-to-end distance in a protein molecule. From the principles of thermodynamics (Treloar 1975), it can be shown that

$$F = \frac{dU}{dr} - T\left(\frac{dS}{dr}\right), \tag{6.51}$$

where U is the internal energy of the system (for our purposes, the energy contained in chemical bonds), T is the absolute temperature, and S is the system's *entropy*, a measure of its disorder.[8]

When the end-to-end distance of a protein molecule changes, it is due almost entirely to the rearrangement of the various links of the polymeric chain, rather than to the stretching or compression of the peptide bonds between amino acids. Thus, we can expect that the change in internal energy with a change in r is small, and we assume here that it is zero. This leads us to conclude that for a protein chain,

$$F = -T\left(\frac{dS}{dr}\right). \tag{6.52}$$

[8] To be completely correct, we should use partial derivatives here to acknowledge the fact that force can vary due to factors other than a change in end-to-end length. But it is only a change in length that we concern ourselves with in the present case, and, as we have done before, we use ordinary derivatives to avoid the complications of introducing new terminology.

In other words, the force required to maintain a given end-to-end distance is proportional to the absolute temperature and how much the disorder (the entropy) of the chain changes as the chain is extended. Because the restoring force in a rubber is a function of the disorder of the rubber's polymer chains, these materials are sometimes referred to as *entropy rubbers.*

Temperature is easily measured, but how can we measure the change in entropy as a polymer chain is stretched? To do so, we borrow a basic relationship from statistical mechanics, in this case Boltzmann's relation that defines entropy:[9]

$$S \equiv k \ln \Omega. \tag{6.53}$$

Here, k is Boltzmann's constant, the same Boltzmann's constant we encountered in chapter 5 when dealing with the motions of gas molecules. (This wonderful number crops up repeatedly in thermal physics and statistical mechanics). Ω is the number of "states" available to a system. In our case, we are concerned with $\Omega(\vec{r})$, the number of different configurations in which the polymer chain can exist while its ends are separated by the distance and direction specified by $\vec{r}$. So, if we know $\Omega(\vec{r})$, we can calculate S. But how can we estimate the number of configurations a chain can take? A bit of mathematical trickery solves the problem.

Let us assume that there is a large number C, the total number of conformations in which our polymer chain can exist for all end-to-end distances. Never mind that we would have a difficult time specifying this number; as we will see, it is sufficient to know that it exists. For any specific separation between the ends of the polymer, only a fraction of this total number of configurations is possible. In fact, we already know what this fraction is. It is simply the probability that the ends of the polymer will be separated by r. In other words,

$$\Omega(r) = C g(r) \, dV, \tag{6.54}$$

where $g(r)$ is the probability density associated with r and dV is again the small bit of volume $4\pi r^2 \, dr$ located at the end of r. Reference to eq. (6.49) thus lets us state that

$$\Omega(r) = C \left(\frac{\sqrt{3}}{\sqrt{\overline{R^2} \sqrt{2\pi}}} \right)^3 \exp\left(-\frac{3r^2}{2\overline{R^2}} \right) dV. \tag{6.55}$$

---

[9] In a sense, this equation also defined Boltzmann; it alone is the epitaph engraved on his tombstone.

With this equation in hand, we are now ready to calculate the entropy S of the polymer chain:

$$S(r) = k \ln \Omega(r)$$

$$= k \ln \left[C \left(\frac{\sqrt{3}}{\sqrt{\overline{R^2}\sqrt{2\pi}}} \right)^3 \exp\left(-\frac{3r^2}{2\overline{R^2}} \right) dV \right]$$

$$= k \ln \left[C \left(\frac{\sqrt{3}}{\sqrt{\overline{R^2}\sqrt{2\pi}}} \right)^3 \right] + k \ln \left[\exp\left(-\frac{3r^2}{2\overline{R^2}} \right) \right] + k \ln(dV)$$

$$= k \ln \left[C \left(\frac{\sqrt{3}}{\sqrt{\overline{R^2}\sqrt{2\pi}}} \right)^3 \right] - k \left(\frac{3r^2}{2\overline{R^2}} \right) + k \ln(dV). \tag{6.56}$$

Recalling that $dV = 4\pi r^2 dr$:

$$S(r) = k \ln \left[C \left(\frac{\sqrt{3}}{\sqrt{\overline{R^2}\sqrt{2\pi}}} \right)^3 \right] - k \left(\frac{3r^2}{2\overline{R^2}} \right) + k \ln\left(4\pi r^2 dr \right). \tag{6.57}$$

This is a nasty-looking expression. But then it isn't really $S(r)$ that we want, it is dS/dr, and taking the derivative of eq. (6.57) with respect to r simplifies matters dramatically. Only the last two terms in eq. (6.57) contain the variable r (as opposed to $\overline{R^2}$, which, as an average, is a constant). Noting that $d[k \ln(4\pi r^2 dr)]/dr = 2k/r$, dS/dr is simply

$$\frac{dS}{dr} = k \left(\frac{-3r}{\overline{R^2}} + \frac{2}{r} \right). \tag{6.58}$$

Recalling from eq. (6.52) that $F = -T dS/dr$, we abruptly find ourselves in a position to calculate the force required to maintain a given r:

$$F = kT \left(\frac{3r}{\overline{R^2}} - \frac{2}{r} \right). \tag{6.59}$$

One last substitution, and we are done. We know that $\overline{R^2} = n\delta^2$, so

$$F = kT \left(\frac{3r}{n\delta^2} - \frac{2}{r} \right). \tag{6.60}$$

This is a result worth pondering. With a little algebra, we can calculate that the force required to keep a polymer chain at a given end-to-end separation is zero when $r = \sqrt{(2n/3)}\delta$. In other words, no force is required to keep the chain's ends at their most probable separation. The more the chain is extended or compressed relative to this most probable separation, the more the force required, just the sort of behavior expected of an elastic material. If an external force is not applied, the chain returns to its most probable (and fairly compact) end-to-end separation. Note that at relatively large values of r, $2/r$ is small compared to $3r/n\delta^2$, and the force required to extend the chain increases approximately linearly with the end-to-end separation. In other words, the chain can be characterized by a more or less constant stiffness.

The relationship we have derived here is for a single protein chain. In a macroscopic piece of rubber, a multitude of these chains is cross-linked to form a three-dimensional network, and it is this network that we think of as "rubber." The bulk properties of the whole network are the same as for the individual chains, however. The sum of elastic proteins is an elastic material.

These results are the end product of a lot of math. Can they be explained in an intuitive fashion? We have already guessed that at its average end-to-end distance, a rubbery protein chain is highly disordered, and it is perhaps intuitive that when the end-to-end distance is increased above the average, the disorder in the chain is reduced. For example, when the chain is extended to its full length, that is, when $r = n\delta$, the chain is stretched straight, a very ordered configuration. Thus, we can deduce that the extension of a protein chain above its average end-to-end length results in a decrease in the disorder of the chain. Because entropy S is a measure of disorder, this means that dS/dr is expected to be negative. Multiplication of this negative number by $-T$ leads us to suppose that a positive force is required to extend the chain.

Examination of eq. (6.60) also shows that the force required to extend a polymer chain increases in proportion to the absolute temperature. This is a phenomenon that can be demonstrated with a rubber band. Stretch a rubber band between your thumbs and heat the center of the band with a hair dryer. As the band heats up, the force required to keep your thumbs a fixed distance apart increases slightly. Another way to think of this property of rubbers is to imagine what would happen to a rubber band if it were heated without having its ends constrained. In this case, the force that would be required to keep the band extended is instead free to cause the band to contract. Indeed, the size of a piece of rubber will decrease as its temperature is raised. This property of entropy rubbers is in distinct contrast to that of other, more typical materials. Steel, for instance, expands when heated. It is this property of rubbers that provided the clue to the structure of spider silk reported in chapter 1.

Equation (6.60) also tells us that the larger the number of links in a polymer chain and the larger the length of each link, the smaller the force required to keep that chain at a given extension. Nature commonly uses twenty different amino acids. Although the side groups of these vary substantially, the distance along the protein chain is set by the sequence of atoms -C-C-N- that is the same for all the common amino acids. Thus, for biological rubbers, which are formed from amino acids, there is virtually no choice in the size of individual links. The number of amino acids per protein is open-ended, however. Thus, it is easy to envision the process by which the stiffness of a biological rubber could be adjusted during the course of evolution. For example, if there were some mechanical advantage to a rubber that was less stiff when used in an insect's wing joint, a mutation that simply added a link or two to each protein chain in the material would give natural selection the necessary variation on which to act.

At this point it is interesting to look back on where we have been. Without rubbery materials, many biological functions would be difficult or impossible. Without a knowledge of probability, it would be difficult or impossible to account for the mechanics of rubbery materials. Thus, the statistics of random walks have led us to the point where we can make productive speculations about the course of evolution in biological structural materials. This is the kind of synthesis of disparate ideas that makes it fun to apply probability and physics in biology.

6.9 *Summary*

In this chapter we have extended our understanding of the biology of random walks beyond the realm of molecular diffusion to include the random motions of plankton and alleles and the three-dimensional statistics of polymeric chains. Although the complexity of these cases has increased relative to those covered in chapter 5, the core message remains much the same. Just as the distance traveled by a thermally agitated molecule depends on the square root of time, the average end-to-end length of a polymeric chain depends on the square root of the number of its links. The more familiar you become with the logical thread that joins these chapters, the better able you will be to recognize similar situations in the real world.

The next chapter will move our discussion in a different direction. Readers who care to pursue the subject of random walks further will find it helpful to consult Berg (1983), Csanady (1973), or Okubo (1980).

6.10 *Problems*

1. The horizontal turbulent diffusivities found in the ocean's surface waters are much larger than the vertical diffusivities found in the same habitat: Mann and Lazier (1991) cite a typical value of $500 \ m^2 \ s^{-1}$. One might assume that given these large diffusivities, planktonic larvae could be effectively

transported throughout the world's oceans. If North America and Europe are thought of as absorbing "walls" that bound the North Atlantic Ocean (approximately 5000 km wide), what is the average time it would take a larva to be delivered to shore by turbulent mixing?

2. The sperm of a hypothetical marine invertebrate moves in a three-dimensional random walk through the water with $\delta = 50$ μm and $\tau = 1$ s. A sperm is released 1 mm away from a spherical egg whose radius is 50 μm. Assume that the sperm swims as if the egg were not present. (This unrealistic assumption means that we won't worry about what happens if the sperm tries to swim through the egg.) Let's further assume that at a particular time the sperm is capable of fertilizing the egg if at that time it is found anywhere in the volume occupied by the egg. Graph the probability of fertilization as a function of time after the sperm's release.

3. Two species of jellyfish are morphologically similar and rely on the elastic properties of their "bells" for locomotion. As a result, both species require a protein rubber that has the same stiffness at high extensions. One species lives in tropical waters, however, where the temperature is 30°C, whereas the other lives in the arctic sea where the temperature is 0°C. How many amino acids would you expect to find in the polypeptide chains of the tropical species' rubbery protein relative to the number found in the arctic species? Express your answer as a percentage.

7

The Statistics of Extremes

In this chapter we follow yet another strange example of a random walk to see where it will lead. Eventually, after wandering through the risks of cocktail parties and ocean waves, we will arrive at a branch of statistics that deals with the extremes of nature, society, and technology. What is the oldest age to which a human being will ever live? Is there ever likely to be another 0.400 hitter in major league baseball? How likely is it that a jet engine will flame out on your next trip to Chicago? When we finish this chapter we will be able to answer these and other such questions.

7.1 *The Danger of Cocktail Parties*

We begin with a historical note. The mathematics we have explored for the last two chapters had its origin not with the consideration of random walks, but rather with a problem in acoustics. John William Strutt (who became Lord Rayleigh upon the death of his father, the baron) was one of the leading applied mathematicians of his day.[1] He had an abiding interest in the physics of sound, and his two-volume treatise on the subject of theoretical acoustics is still in print (Rayleigh 1945).[2] In the course of exploring the many ramifications of sound, Rayleigh considered what could be called the "Cocktail Party Problem."

Imagine yourself at a cocktail party. If the party is typical of such festivities, you find yourself confined to a small room with fifty other revelers, each of whom is busily engaged in a shouted conversation with a neighbor. As anyone who has had the misfortune of being subjected to this torture can attest, the level of sound in such a party can be annoying. Therein lies the problem; a simple analysis suggests that the sound in a cocktail party could rise to the point where it could literally be deafening. The reasoning goes as follows.

---

[1] Among many other accomplishments, he was awarded the Nobel Prize in physics in 1904 for his role in the discovery of the element argon.

[2] In true Victorian fashion, this monumental work was written in part while Rayleigh was traveling up the Nile on a houseboat. He was recovering from a bout of rheumatic fever at the time.

Sound is a wave of oscillating pressure that travels through a fluid. That is, if you measure the pressure in the air around you, it rises and falls as waves of sound move past, sometimes being higher than the average, sometimes lower. A simple, pure tone is a sinusoidal fluctuation in pressure as depicted by any of the dashed curves in figure 7.1. This type of variation in pressure can be represented mathematically as

$$\mathcal{P}(t) = \mathcal{P}_{amb} + a \sin(2\pi f t + \phi), \tag{7.1}$$

where $\mathcal{P}(t)$ is the instantaneous pressure at time t, and $\mathcal{P}_{amb}$ is the ambient atmospheric pressure. The magnitude of the deviation in pressure is set by the wave amplitude, a. How rapidly the pressure varies through time is set by the frequency f, and the time at which the pressure reaches its maximum is set by the phase coefficient ϕ. The sum $(2\pi f t + \phi)$ is the *phase* of the wave.

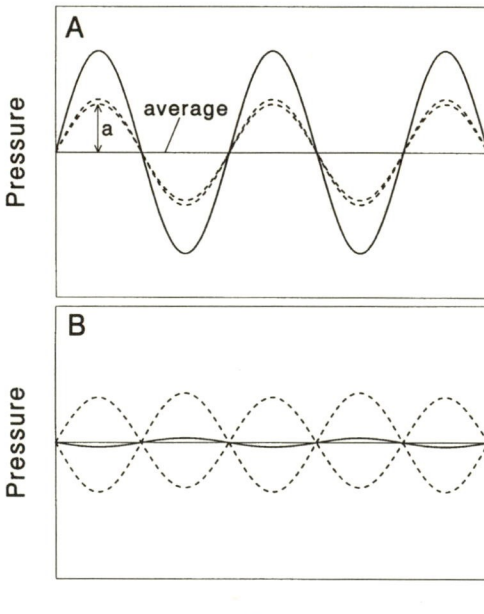

FIG. 7.1 The pressures due to sound are additive. For example, when two waves are in phase (A), the resulting net variation in pressure (the solid curve) is larger than either component (dashed curves). When waves are out of phase (dashed lines in B), the resulting net variation in pressure (the solid line) can be small.

To a good approximation, sound waves are additive. If two sound waves are produced such that the peak positive pressure of one arrives at your ear at the same time as the peak positive pressure of the other (for example, the two waves shown by dashed lines in fig. 7.1A), the net pressure imposed on your eardrum is the sum of the pressures from the two waves—its amplitude is increased (as shown by the solid line in fig. 7.1A). In this case, the waves are said to be "in phase." By the same token, if the peak positive pressure of one wave arrives at the same time as the peak negative pressure from the other, the two waves are maximally out of phase, the two pressures tend to offset, and your eardrum is subjected to a much smaller variation in pressure. This effect is shown in figure 7.1B. Here one of the constituent waves from figure 7.1A has been shifted so that it is maximally out of phase with the other; the net result is that pressure varies much less.

Now consider a situation in which a large number of sound sources are present. Each source produces sound with the same amplitude of pressure, but with a slightly different frequency and a randomly determined phase coefficient. The slight difference in frequency results in a continuously changing amplitude at your eardrum as waves come in and out of phase (fig. 7.2). The random choice of phase coefficients means that we cannot predict exactly what the "mix" of sounds will be at any given time, and therefore cannot predict the precise shape of the wave form at your ear. The situation described here is a reasonable approximation of the sound at a cocktail party. True, when people talk they do not produce pure tones, but the range of frequencies used in human speech is relatively small. People at a party certainly do not speak in unison, ensuring that the phases of the sounds produced will be unpredictable when they reach your ears.

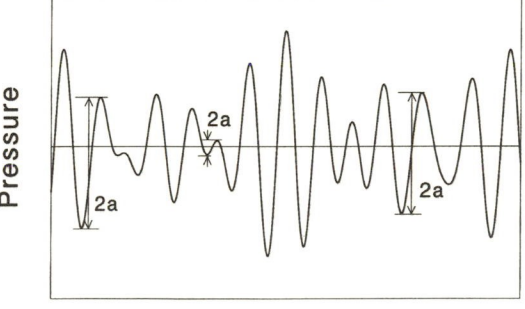

Fig. 7.2 The sum of multiple waves with similar periods and random phases produces a randomly varying waveform. The amplitude *a* is defined as half the pressure difference between a wave peak and the preceding minimum.

Given this background, we can now pose the Cocktail Party Problem as a more formal question. In a given period, what is the maximum amplitude of sound that is expected to arrive at your ear? It is easy to imagine that if there are fifty people at a party, the sound waves from every one could arrive at your ear in phase, painfully subjecting your ear drum to a pressure fifty times the pressure due to a single person's shout. How likely is it that this dire scenario will occur?

The path leading to an answer becomes evident when we realize that the sound pressure at any time is the sum of a number of individual pressure deviations with random magnitudes. This is directly analogous to the displacement of a particle undergoing a random walk in three dimensions—at any time the displacement of the particle along one axis is the sum of a number of steps, each with a random length.[3] In other words, we can think of the variation of the sound pressure from one peak to the next as a random walk along an axis representing pressure amplitude.

This was the insight of Lord Rayleigh. When he worked through the math, he found that the probability density of the net pressure amplitude at your ear is

$$g(A) = \frac{2a}{a_{\text{rms}}^2} \exp\left[-\left(\frac{a}{a_{\text{rms}}}\right)^2\right], \qquad (7.2)$$

a distribution now known as the *Rayleigh distribution* (Rayleigh 1880). Here a is the amplitude of a single peak in sound pressures one realization of the random variable, A. To be precise, a is equal to half the change in pressure between one pressure "crest" and the preceding pressure "trough" (see fig. 7.2). a_{rms} is the root mean square amplitude of the sound arriving at your ears:

$$a_{\text{rms}} = \sqrt{\frac{1}{n}\sum_{i=1}^{n} a_i^2}, \qquad (7.3)$$

where each a_i is an individual amplitude and n is the number of amplitudes in the sample.[4]

As you might reasonably expect, the shape of the Rayleigh distribution is similar to the distribution we obtained for the end-to-end distance of a polymer chain (fig. 7.3). The probability of having a net amplitude near zero is low, because that would require the unlikely circumstance that numerous waves combine by chance to just offset one another. The probability density curve rises to

---

[3] The step length taken by the particle may be fixed, as we assumed in chapter 5. However, because the particle steps in random directions, the projection of this fixed length step on any axis has randomly varying lengths.

[4] Note that a_{rms} is different from σ, the standard deviation of pressure. To measure the standard deviation in pressure, you would record the pressure at a series of times and calculate first the average pressure and then the root mean square deviation from this average. This rms deviation is the standard deviation. In contrast, a_{rms} is calculated using only the *peaks* in pressure, and consequently $a_{\text{rms}} > \sigma$. To a close approximation $a_{\text{rms}} = \sqrt{2}\sigma$.

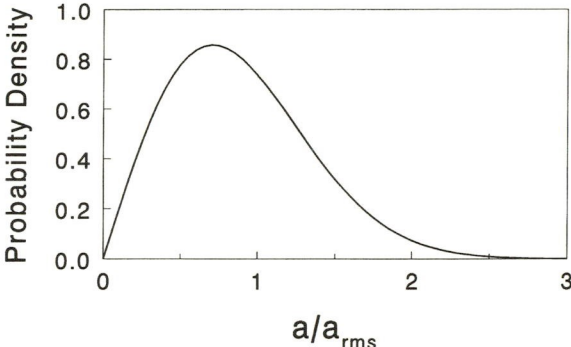

FIG. 7.3 The Rayleigh probability density function of pressure amplitudes (eq. 7.2). The long tail to the right is the root of the worry in attending a cocktail party. Note that values on the x-axis have been normalized to the root mean square amplitude, an index of the average sound level.

a maximum at $a = (1/\sqrt{2})a_{rms}$; that is, the most probable amplitude is approximately 0.71 times the root mean square amplitude. The probability density curve has a long tail extending off toward very high amplitudes, corresponding to the slight probability that numerous waves come into phase at your ear. It is this tail that is of primary interest in the context of the Cocktail Party Problem, because it is there that the potentially damaging, large-amplitude sounds are found.

Now, the Rayleigh probability density function expresses the chance that an amplitude chosen at random lies within a certain range. Of greater concern in the present context is the chance that an amplitude chosen at random will exceed the pressure required to damage your ears. To calculate this probability we use a common trick of probability theory (see chapter 2) and calculate the probability that one peak chosen at random will have *less* than a given amplitude. This is easily done. Recall from chapter 4 that a probability density curve is simply the derivative of its corresponding cumulative probability curve. Thus, by integrating $g(A)$ with respect to a, we can calculate the cumulative probability, in this case the chance that a wave peak chosen at random will have less than a given amplitude. First, we note that

$$g(A) = \frac{2a}{a_{rms}^2} \exp\left[-\left(\frac{a}{a_{rms}}\right)^2\right] = \frac{-d\exp\left[-\left(a/a_{rms}\right)^2\right]}{da}. \qquad (7.4)$$

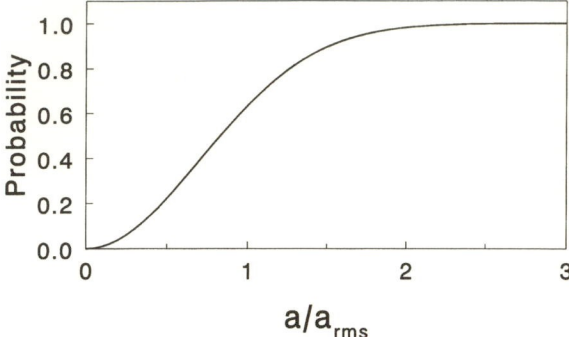

FIG. 7.4 The cumulative probability distribution for a
Rayleigh probability density function (eq. 7.5). The
larger the specified amplitude, the higher the probability
that a wave peak chosen at random will have less than
that amplitude. Note that values on the x-axis have been
normalized to the root mean square amplitude, an index
of the average sound level.

Thus,

$$P(A \leq a) = \int_0^a g(A)da$$

$$= -\int_0^a d \exp\left[-\left(\frac{a}{a_{rms}}\right)^2\right]$$

$$= 1 - \exp\left[-\left(\frac{a}{a_{rms}}\right)^2\right]. \tag{7.5}$$

This cumulative probability function is graphed in figure 7.4.

Let's pause for a moment to consider the meaning of this curve. If we know
the average amplitude of sound impinging on your ears (that is, a_{rms}), we can
specify the probability that any amplitude that we choose at random is less
than a given value, a. For example, if the "average" sound level at our cocktail
party is taken to be a_{rms}, the probability that a wave peak chosen at random
has an amplitude less than average is $1 - e^{-1} \cong 0.63$. In other words, there is
a 63% chance that a randomly chosen wave amplitude is less than the average
amplitude of sound in the room.

Now, if the probability expressed by eq. (7.5) is the probability that a randomly chosen amplitude is less than or equal to a, 1 minus this value is the

probability that a randomly chosen amplitude a_i is *greater* than a:

$$P(a > A) = 1 - \left\{ 1 - \exp\left[-\left(\frac{a}{a_{\text{rms}}}\right)^2 \right] \right\}$$

$$= \exp\left[-\left(\frac{a}{a_{\text{rms}}}\right)^2 \right]. \tag{7.6}$$

For example, if $a = 2a_{\text{rms}}$, the probability that a wave will exceed a is $e^{-4} = 0.018$. Fewer than two waves in a hundred have an amplitude twice the average.

The probability of exceedance that we have just calculated concerns one wave peak chosen at random. But one is seldom so lucky as to be able to duck out of a cocktail party after being exposed to a single peak of sound pressure. The longer you are at the party, the more sound waves reach your ear, and the greater the likelihood that one of these wave peaks will do you damage. How can we relate the probability of exceedance for one wave to the risk associated with many waves?

We begin by introducing the concept of the expected exceedance number, N_x. If the probability of randomly encountering a sound wave whose amplitude a_i is greater than a is 1/10, we have shown in chapter 3 that we would expect to sample approximately ten waves before encountering one for which $A \geq a$. Thus, the expected number of waves you would have to sample to exceed a given amplitude is

$$N_x = \frac{1}{P(A \geq a)} = \frac{1}{\exp\left[-(a/a_{\text{rms}})^2\right]}. \tag{7.7}$$

If sound waves have frequency f (measured in waves per second, Hz), the wave period (the time between the arrival of successive wave crests) is $1/f$. As a consequence, the time corresponding to N_x is the *return time*, t_r:

$$t_r = \frac{N_x}{f}, \tag{7.8}$$

where t_r in this case is measured in seconds. The return time is thus the time we would expect to wait between the arrival of waves that exceed a. Dividing N_x by f in eq. (7.7) we find that

$$t_r = \frac{1}{f \exp\left[-(a/a_{\text{rms}})^2\right]}. \tag{7.9}$$

Herein lies the first hint as to why your ears aren't routinely damaged at cocktail parties. For example, let us assume that 500 Hz is a representative frequency for human speech. If the level of sound required to damage your ears is five times the average sound level at a party, we calculate that the return time for $a = 5a_{rms}$ is 4.6 years. In other words, you would have to attend a cocktail party continuously for more than five and a half years before it was an even bet that your ears would be damaged. If the critical level of sound is ten times the average, a much longer period would be required (1.7×10^{33} years!).

7.2 *Calculating the Maximum*

So far we have dealt only with the probability that a sound wave will exceed a given amplitude. We have not, however, calculated exactly how large the largest pressure amplitude will be. For instance, if you are stuck in a perpetual cocktail party for 4.6 years, you can expect the sound to exceed five times the average, but at this point we can't say by how much the sound will exceed this level. Is it possible to be more specific as to what the maximum amplitude actually is? To do so we need to examine in detail the distribution of extreme sound pressures.

Consider a sample of n amplitudes. We have already calculated the probability that any particular wave amplitude is less than a:

$$P(A \le a) = 1 - \exp\left[-\left(\frac{a}{a_{rms}}\right)^2\right]. \tag{7.10}$$

To streamline our notation, let's define a new symbol, θ:

$$\theta(a) = \exp\left[-\left(\frac{a}{a_{rms}}\right)^2\right]. \tag{7.11}$$

Thus,

$$P(A \le a) = 1 - \theta(a). \tag{7.12}$$

If the amplitude of each peak is independent of the amplitude of other peaks, the chance that *every* amplitude in a sample of n amplitudes is less than a is the product of the probabilities of individual amplitudes, or

$$P(\text{all amplitudes} \le a) = [1 - \theta(a)]^n. \tag{7.13}$$

If we know the probability that all n peaks have amplitude *less than or equal to a*, the probability that at least one peak has an amplitude *greater* than a is

$$P(\text{at least one amplitude} > a) = 1 - [1 - \theta(a)]^n. \tag{7.14}$$

This function is shown in figure 7.5. It is absolutely certain that at least one peak will have an amplitude greater than zero, and the probability of exceedance

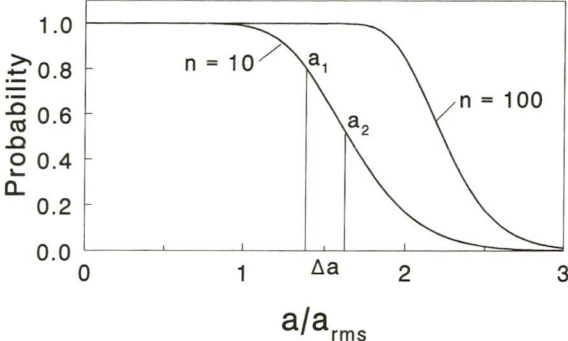

FIG. 7.5 The probability that at least one wave amplitude in a sample of n amplitudes will exceed a given value (eq. 7.14). The larger the number of waves sampled, the greater the probability of encountering a wave with large amplitude. Note that values on the x-axis have been normalized to the root mean square amplitude, an index of the average sound level.

decreases rapidly as a increases beyond a certain critical value. The more peaks in our sample, the larger the largest amplitude we can expect, and the higher the critical value of a. Examples are shown here for samples with ten and hundred waves.

Now the real fun begins. Consider one small segment of the exceedance curve in figure 7.5. The probability of exceedance is higher at point a_1 than it is a small distance Δa to the right at point a_2. The difference between these two probabilities is the likelihood that at least one peak has an amplitude greater than a_1 but that the amplitude of none of these peaks is greater than a_2. In other words, the difference in probability between points a_1 and a_2 is the probability that the maximum amplitude—a_{max} in our sample of n peaks—lies between a and $a + \Delta a$. Thus, by calculating how fast this probability changes as a function of a, we can calculate the probability that the random variable A_{max} (of which a_{max} is an individual realization) falls within a given range. But the probability that A_{max} lies within a specified range is nothing other than the probability density of A_{max}. This is progress!

Putting these thoughts into an equation, we see that

$$g(A_{max}) = \frac{\{1 - [1 - \theta(a)]^n\} - \{1 - [1 - \theta(a + \Delta a)]^n\}}{\Delta a}. \tag{7.15}$$

By the definition of a derivative, as Δa goes to 0,

$$
\begin{aligned}
g(A_{\max}) &= \frac{-d\left[1 - [1 - \theta(a)]^n\right]}{da} \\
&= \frac{d\left[1 - \theta(a)\right]^n}{da} \\
&= n\left[1 - \theta(a)\right]^{n-1}\left[\frac{-d\theta(a)}{da}\right].
\end{aligned}
\tag{7.16}
$$

From the definition of $\theta(a)$ we know that

$$
\begin{aligned}
\frac{-d\theta(a)}{da} &= \frac{-d\left\{\exp\left[-\left(a/a_{\mathrm{rms}}\right)^2\right]\right\}}{da} \\
&= \frac{2a}{a_{\mathrm{rms}}^2}\exp\left[-\left(\frac{a}{a_{\mathrm{rms}}}\right)^2\right] \\
&= g(A).
\end{aligned}
\tag{7.17}
$$

So,

$$
g(A_{\max}) = n\left[1 - \theta(a)\right]^{n-1} g(A).
\tag{7.18}
$$

This, believe it or not, is the answer we seek. The probability density of finding the maximum amplitude $A_{\max}$ in a given range of A is equal to $g(A)$, the probability of finding an individual amplitude in that range, tempered by the term $n\left[1 - \theta(a)\right]^{n-1}$. This term is small when a is small, whereas $g(A)$ is small when a is large. The net result is that $g(A_{\max})$ has an appreciable value only for a narrow range of amplitude. Expanding all the terms in eq. (7.18),

$$
g(A_{\max}) = n\left\{1 - \exp\left[-\left(\frac{a}{a_{\mathrm{rms}}}\right)^2\right]\right\}^{n-1}\left(\frac{2a}{a_{\mathrm{rms}}^2}\right)\exp\left[-\left(\frac{a}{a_{\mathrm{rms}}}\right)^2\right],
\tag{7.19}
$$

and plotting the results (fig. 7.6), we see what this probability distribution of maximum amplitudes looks like. The largest value in $g(A_{\max})$ the *mode*, corresponds to the steepest dropoff in the curve of $P(A_{\max} > a)$, and the location of the mode of $g(A_{\max})$ varies with n, the number of waves encountered. The larger n is, the higher the pressure amplitude at which the mode occurs and the narrower the distribution.

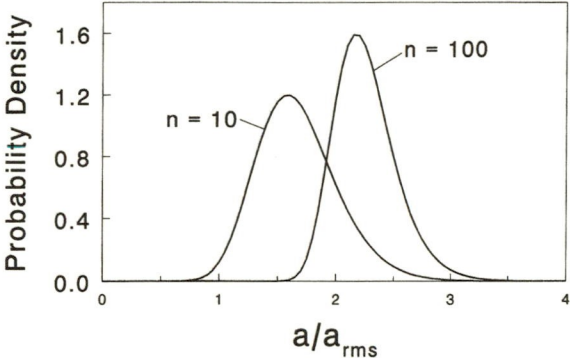

FIG. 7.6 The probability density function for the *maximal* amplitude in a sample of *n* amplitudes (eq. 7.19). The larger the number of waves in a sample, the larger the maximum value expected and the narrower the distribution of maximal amplitudes.

7.3 *Mean and Modal Maxima*

Our calculations to this point have allowed us to specify that the maximal sound pressure imposed on your ears is likely to fall within a narrow range as defined by the probability density distribution (eq. 7.19). It would be convenient to characterize this distribution in the same fashion as we characterized the distribution of end-to-end distances in a polymer chain, that is, by its mean and its mode. This can be done, but the mathematics is tortuous, so only the final results are presented here. In a seminal paper published in 1952, Michael Longuet-Higgins showed that the mode of $g(A_{max})$, that is, the most probable maximum amplitude, is

$$\text{modal } a_{max} \cong a_{rms} \sqrt{\ln(n)}, \qquad (7.20)$$

where n is the number of peaks in the sample. Note that this is an approximation. The value calculated here is slightly smaller than the actual value, but the error is only approximately $1/[\ln(n)]^{3/2}$. When n is large, this error is quite small, and this estimate of the mode is accurate.

Because the probability density distribution of a_{max} has a longer tail for high a than for low a (fig. 7.6), the mean a_{max} is greater than the mode. Longuet-Higgins showed that

$$\text{mean } a_{max} \cong a_{rms} \left(\sqrt{\ln(n)} + \frac{\gamma}{2\sqrt{\ln(n)}} \right). \qquad (7.21)$$

Again, there is an error on the order of $1/[\ln(n)]^{3/2}$. The constant γ is Euler's constant, $0.5772\ldots$.

Eqs. (7.20) and (7.21) are expressed in terms of n, the number of waves encountered. We can easily convert n to time t by noting that $n = ft$, where f is again the sound's frequency. Thus,

$$\text{modal } a_{\max} \cong a_{\mathrm{rms}}\sqrt{\ln(ft)}. \tag{7.22}$$

$$\text{mean } a_{\max} \cong a_{\mathrm{rms}}\left(\sqrt{\ln(ft)} + \frac{\gamma}{2\sqrt{\ln(ft)}}\right). \tag{7.23}$$

These expressions are graphed in figure 7.7, in which we have again assumed that human speech can be characterized by a frequency of 500 Hz. We see that the expected (mean) maximal sound amplitude you will encounter in one second is 2.6 times the root mean square amplitude. In one minute, the expected maximum has increased to 3.3 times a_{rms}, but in one hour the further increase is slight and the expected maximum is 3.9 a_{rms}. Thus, if you can stand the noise for one minute, you are probably safe in staying for an hour.

This, then, is the final answer to the Cocktail Party Problem. Yes, there is indeed a probability that sound from every party goer will reach your ear in phase, but in any reasonable period the chance is negligibly slight. Victorian society must have been quite relieved by Lord Rayleigh's assurance that cocktail parties were unlikely to lead to sonic doom. Perhaps this is why the tradition of hosting such parties has survived to this day.

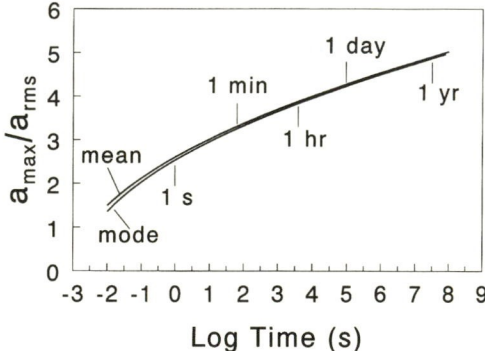

FIG. 7.7 The mean (expected) and modal (most probable) maximum amplitudes vary with the time to which your ear is exposed to sound (eqs. 7.23 and 7.22, respectively). The curves shown here assume that human speech has a frequency of about 500 Hz. Note that values on the y-axis have been normalized to the root mean square amplitude, an index of the average sound level.

7.4 Ocean Waves

Cocktail parties are not the only occasions on which these calculations can be of use. A practical, biological example concerns the forces exerted on intertidal organisms. In this case, the waves we deal with are not sound waves but rather ocean waves, and the amplitude of the wave is measured in terms of vertical deviation from mean sea level rather than deviations from atmospheric pressure. Otherwise, the mathematics are the same.

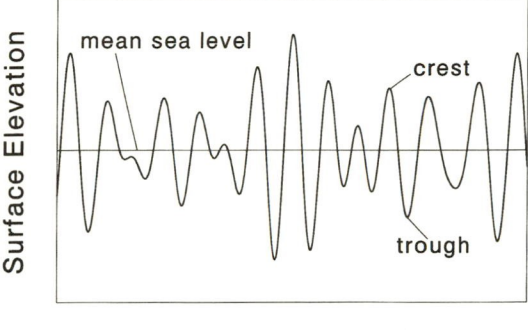

FIG. 7.8 A hypothetical trace of sea-surface elevation as a function of time. The similarity between this curve and that of figure 7.2 is intentional; the mathematics used to describe sound waves can also be used for ocean waves.

To see how this works, consider the example shown in figure 7.8, a hypothetical record of sea-surface elevation through time. The similarity between this wave form and that of figure 7.2 is intentional; the chance combination of ocean waves leads to the same sort of modulation in amplitude that we encountered with pressure waves. In the real world, a record such as this could be obtained in a number of ways. For example, a buoy floating on the ocean's surface moves up and down as waves pass by, and a force transducer mounted in the buoy can sense the accelerations associated with this motion. Integrating these accelerations through time yields the vertical velocity of the buoy, and integrating these velocities provides a measure of the buoy's (and hence the water surface's) vertical location. Alternatively, surface elevation can be calculated from a record of the subsurface hydrostatic pressure. As wave crests and troughs pass over a submerged pressure transducer, the pressure increases and decreases, providing the necessary information. For a more thorough discussion of the mechanics of measuring ocean waves, consult Denny (1988). For the moment, let's concentrate on how to interpret this record.

Each wave has a trough and a crest, and it is possible (if laborious) to measure the vertical distance from each crest to its preceding trough. Dividing each of these wave heights by 2 gives us the wave amplitudes we use in our analysis. The root mean square amplitude is then calculated using eq. (7.3). This is a measure of the "waviness" of the ocean's surface, and as we have seen, it forms an integral part of the calculation of maximal wave amplitude.

From our wave record we can also measure the interval between the arrival of wave crests, and from these data we can calculate the mean wave period, $\mathcal{T}$. The wave frequency, f, is simply $1/\mathcal{T}$.

With a measure of a_{rms} and f now in hand, we are free to use eq. (7.23) to calculate the maximal wave amplitude an intertidal organism would expect to encounter in a given period. For example, if the root mean square amplitude is 1 m (a common value for moderately exposed shores) and the average wave frequency is 0.1 Hz, the largest wave amplitude encountered in a 24-hour period is expected to be 3.1 m.

The value of this computation lies in the relationship between the wave amplitude and the water velocity imposed on an organism. To a rough approximation, the maximal water velocity associated with a wave as it breaks is

$$u_{max} = \sqrt{g(2a_{max} + d)}, \tag{7.24}$$

where g is the acceleration due to gravity (9.81 m s^{-2}), and d is the depth of the water column under the breaking wave. For example, if the 3.1 m amplitude peak noted above were to break in water that was 6.2 m deep (as it probably would in reality), the maximal velocity in the wave would be about 11 m s^{-1}.

If this wave breaks on a steep shore, this maximal velocity is maintained as the wave crest arches over and strikes the shore. The resulting flow can impose both a drag and a lift force on organisms attached to the surf-zone rocks. In many cases, drag, a force in the direction of flow, is the larger of these two forces, and we use it as an example here. For the rapid water velocities found in the surf zone, the drag on an object can be accurately predicted using the following relationship (Vogel 1994; Denny 1988):

$$\text{drag} = \frac{1}{2}\rho u_{max}^2 A C_d. \tag{7.25}$$

Here ρ is the density of seawater (1025 kg m^{-3}), A is the area of the organism exposed to flow (usually measured as the area projected in the direction of flow), and C_d is the drag coefficient, an index of the organism's shape.

To give this equation some tangibility, let's calculate the drag force you might personally experience if exposed to a wave with an amplitude of 3.1 m. If you were to stand upright on the rock and let the wave impact you frontally (not a good idea), the projected area you would expose to flow would be approximately 1 m^2. The drag coefficient for a human being in this orientation is about 1.0 (Hoerner 1965). As we calculated above, the maximum velocity associated with a wave amplitude of 3.1 m is 11 m s^{-1}. Chugging through the numbers, we find that you would expect to feel a drag of 6.2×10^4 newtons. This is equivalent to the weight of 6.9 tons, the equivalent of two medium-size elephants! If you ever wondered (as we did as children) why the victims of shipwrecks didn't just swim ashore, here is the answer. There are exceptionally large forces associated with breaking waves—more than sufficient to maim an unlucky sailor—and these forces have served as selective factors in the evolution of size and form in wave-swept organisms. For a more thorough discussion of the statistics and biology of ocean waves, the interested reader should consult Denny (1988, 1995).

7.5 *The Statistics of Extremes*

The examples we have explored in this chapter (waves of sound and water) are just two of the many situations in which it is the extreme of a property rather than its average that has practical importance. In the context of a cocktail party, you are likely to care more about the maximal level of noise than about the average level. A barnacle on an intertidal rock is likely to have its life affected more by the maximal water velocity it encounters than by the average. Similarly, the extremes in temperature may have more drastic biological effects than the typical temperatures encountered every day. If the temperature gets too high or too low, a given plant or animal will die. Other examples abound: extremes in rainfall, light intensity, salinity, wind speed—all have important biological consequences.

The calculations we have made concerning maximum wave amplitudes provide an example of how we can predict the extremes of a random process. But waves (both sound and ocean) are unusual in that we know from first principles something about their probability distributions. What if we want to calculate an extreme temperature? There is no thermal equivalent to the Rayleigh distribution that will tell us a priori how temperatures fluctuate through time. The same is true for rainfall, wind speed, and many other examples of environmental fluctuations. If we do not know the probability distribution of individual fluctuations, how can we make predictions about the extremes? Fortunately, a method exists, and it actually is quite straightforward. It even has a straightforward name —the *statistics of extremes*.

We begin by asking you to remember our discussion of probability density functions in chapter 4. There we noted that means drawn from a given population are normally distributed even if the population itself is not. In other words, the distribution of means is conveniently independent of the distribution of the parent population. In this case, the independence is embodied in the Central Limit Theorem, and it forms the basis for much of inferential statistics.

An analogous property holds in relation to the extremes (rather than the means) sampled from a population: the distribution of the extreme values of samples drawn from a population can often be predicted independent of the distribution of the parent population. In this case, the extremes do not all conform to a single probability distribution (as do the means). Instead, for a wide variety of parent distributions, the extremes conform to one of three probability distributions. These three *asymptotic extreme value distributions* were popularized by E. J. Gumbel in 1958 and are often referred to as Gumbel types I, II, and III, respectively:

$$P(x_{max}) = \exp\{-\exp[k_1(x_{max} - k_2)]\},$$ (7.26)

or

$$P(x_{max}) = \exp\left[-\left(\frac{b_1}{x_{max} - b_2}\right)^{b_3}\right],$$ (7.27)

or

$$P(x_{max}) = \exp\left[-\left(\frac{c_1 - x_{max}}{c_1 - c_2}\right)^{c_3}\right].$$ (7.28)

In each case, $P(x_{max})$ is the probability that an individual extreme value (chosen at random) is less than or equal to x_{max}.

An example may be in order. The extremes of samples taken from a normal distribution have a Gumbel Type I cumulative probability distribution. So do the extremes sampled from an exponential distribution and the extremes of samples from a Poisson distribution. In other words, many different probability distributions can provide samples whose extreme values have the same probability structure.

This simplifies our task considerably. If we are trying to predict the properties of the extremes of a given variable (such as temperature), we do not necessarily need to know the details of the distribution of the variable itself. All we really need to know is which of these three asymptotic distributions captures the behavior of its extremes. But, without knowing the distribution of the variable itself, how can we know which asymptotic distribution is appropriate? Again, statistics comes to the rescue. It is possible to proceed from empirical data to arrive at the appropriate asymptotic extreme value distribution. Here is how it works.

Let's assume that we want to explore the extreme values of a process that varies randomly through time, but the same ideas can be applied to processes that vary through space or any other dimension. To keep matters simple, we specify that our variable (temperature, for instance) fluctuates randomly around a mean value, but that the mean value is constant for all time. In the jargon of statistics, we are thus dealing with a *stationary* process. Furthermore, we assume that the process that controls the fluctuation of our variable does not change through time; that is, that the variation is *ergodic*. An example of such a process is shown in figure 7.9, one portion of a hypothetical time series of hourly temperatures. The mean temperature here is 10°C, and the daily variation in temperature from night to day is evident. There are, however, random fluctuations that modify this predictable variation.

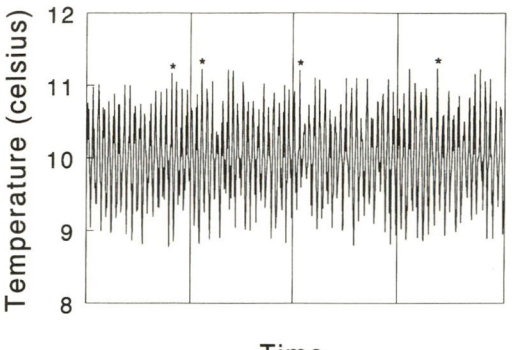

FIG. 7.9 A portion of a hypothetical time series of temperatures. The starred peaks are the maximal temperatures in each of the four 20-day intervals.

Our first step is to divide our record of temperature into a series of intervals of equal length, as shown by the vertical lines in figure 7.9. The intervals need not be contiguous, but in practice they usually are. Why waste information by leaving gaps?[5] Within certain limits, the length we choose for our intervals is not important. The main criterion is that each interval should contain a sufficient number of fluctuations so that we can pick an "honest" maximum. For example, it would be possible to pick such a short interval that each contained only one fluctuation, and the peak value in this fluctuation would be the "extreme" value for the interval. But there really isn't anything extreme about this value; it's just a local blip. A rough rule of thumb is that each interval should contain at least 20–30 fluctuations.

On the other hand, we could set the interval size equal to the entire length of our temperature record. In this case we would be assured of picking an honest extreme, but we would only have a single sample (one interval) to work with. In practice, the interval size should be set short enough so that twenty or more sample extremes can be measured. In the example shown in figure 7.9, we have chosen an interval of twenty days.

Having picked an interval size, we next note the extreme value found in each interval, as shown by the starred peaks in figure 7.9. If we have n intervals in the entire time series, we thus end up with n estimates of the extreme temperature. For simplicity here, we will only discuss the maximal temperatures, but the same logic can be applied to minimal temperatures as well.

[5] On the other hand, if you have gaps in your data (you dropped the thermometer), it poses no problem.

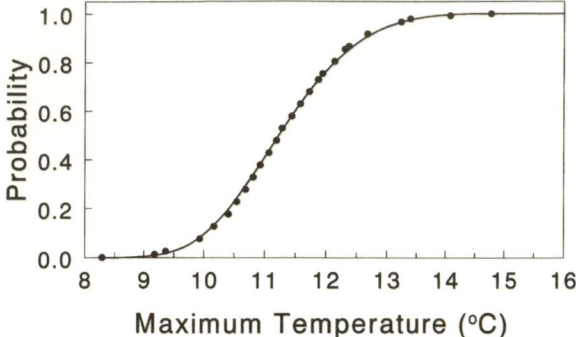

F<small>IG</small>. 7.10 A hypothetical cumulative probability curve of
maximal temperatures. The dots represent the measured
data, and the solid line is the theoretical curve (eq. 7.30)
fitted to the data.

The next step is to estimate the cumulative probability distribution of these
n extremes. This is accomplished by ranking the sample extremes in ascending
order. The smallest maximum temperature thus has rank 1, the next smallest,
rank 2, and the largest extreme temperature, rank n. These ranks can then be
used to provide an estimate that a sample extreme (taken from a randomly
chosen interval) would be less than a given temperature. Gumbel (1958) has
shown that the best estimate of the cumulative probability associated with each
of our ranked sample extremes is

$$P(x_{max} \leq x_i) = \frac{i}{n+1}. \tag{7.29}$$

Here x_i is the magnitude of the extreme of rank i. In our example, x_1 is the
lowest extreme temperature and x_n is the highest extreme temperature found
in our n intervals. In the case where two (or more) extremes have the same
magnitude, they are assigned an average rank.

A graph of cumulative probability as a function of the magnitude of the
extreme shows what data we have up to this point (fig. 7.10). For this example of
temperature, we find that there is a 50% estimated probability that the maximal
temperature in an interval chosen at random is less than or equal to 11.2°C
above the mean temperature. This empirical cumulative probability curve thus
provides us with a rough guess as to the probability structure of the extremes
of temperature in our system.

But we can go farther! We know that if we had a very long temperature
record (and, therefore, a very large number of sample extremes), our empirically determined cumulative probability curve would likely asymptote to one of
Gumbel's three types. Knowing this, we need only collect enough data to give

us a reliable guess as to which type is appropriate and what values to use for the coefficients in the asymptotic distribution. In practice, as few as twenty sample extremes may be sufficient.

The choice among types and the adjustment of the coefficients is accomplished with the help of any standard computerized statistical package. The non-linear curve-fitting module of the program is used to fit a generalized cumulative probability equation to an empirically determined probability function. This expression incorporates all three of the Gumbel curves:

$$P(x_{\max} \leq x) = \exp\left[-\left(\frac{\alpha - \beta x}{\alpha - \beta \epsilon}\right)^{1/\beta}\right], \tag{7.30}$$

with the following provisos:

$$\text{if } \beta > 0, \quad P = 1 \quad \text{for } x \geq \alpha/\beta, \tag{7.31}$$

$$\text{if } \beta < 0, \quad P = 0 \quad \text{for } x \leq \alpha/\beta. \tag{7.32}$$

In an iterative process, the program determines what values of α, β, and ϵ give the best fit to the empirical data. The sign of β then provides the best guess as to which of Gumbel's three asymptotic equations you are dealing with:

- If $\beta \approx 0$ (in practice, if $|\beta| < 0.01$ or so), you have a Type I distribution with $k_1 = 1/\alpha$ and $k_2 = \epsilon$.
- If $\beta < 0$, you have a Type II distribution, with $b_1 = \epsilon - \alpha/\beta$, $b_2 = \alpha/\beta$, and $b_3 = \beta$.
- If $\beta > 0$, you have a Type III distribution with $c_1 = \alpha/\beta$, $c_2 = \epsilon$, and $c_3 = 1/\beta$.

With the coefficients α, β, and ϵ in hand, it is an easy matter to calculate values of biological interest. For example, we might know from laboratory experiments that acorn barnacles die if the temperature exceeds T_{crit}. How often is this temperature reached on the intertidal rocks where these animals live? To answer this question, we record the maximum temperature on a rock for each of a series of 20-day intervals. By fitting these data to eq. (7.30) we can calculate from the probability $P(T_{\max} \leq T_{\text{crit}})$ that in a 20-day interval chosen at random the maximum temperature $T_{\max}$ will be less than T_{crit}. From this probability we can then calculate a return time:

$$t_r = \frac{\text{interval length}}{1 - P(T_{\max} \leq T_{\text{crit}})}. \tag{7.33}$$

Note that if the interval length is measured in days, t_r will likewise have units of days. If this return time is less than the time required for a barnacle to reach sexual maturity, it is unlikely that acorn barnacles can successfully colonize that particular rock. If, on the other hand, t_r is substantially greater than the time to sexual maturity, barnacles may be able to reproduce successfully, although they will occasionally be killed by rare hot days.

7.6 *Life and Death in Rhode Island*

The statistics of extremes can be used in a wide variety of real-world situations. Just about any process that incorporates an element of chance and allows for the measurement of maximum or minimum values is fair game for this kind of scrutiny. Consider the human life span, for instance. How long you are likely to live certainly depends in part on luck, but with the advent of new techniques for fighting disease and replacing worn-out joints and organs, a larger fraction of people are living to old age. Can this "progress" continue forever, or is there some well-defined upper limit to our life span? The statistics of extremes can help to provide an answer

We begin with a reconsideration of eq. (7.30):

$$P(x_{\max} \leq x) = \exp\left[-\left(\frac{\alpha - \beta x}{\alpha - \beta \epsilon}\right)^{1/\beta}\right].$$

If we can find a value of x such that the term in the square brackets is 0, the probability that $x_{\max} \leq x$ is 1 (because $e^{-0} = 1$). In other words, for x such that the term in square brackets is zero, it is absolutely certain that all values encountered at random are less than or equal to x, and there consequently exists an absolute upper limit to the maximum value that can occur. Under what conditions can the term in square brackets be zero? First, β must be positive (indicating a Gumbel Type III curve); if $\beta \leq 0$ there is no meaningful way in which the term in square brackets can be 0. However, if $\beta > 0$, the whole term in square brackets is equal to 0 when $x = \alpha/\beta$. That is, if $\beta > 0$, the absolute largest value of $x_{\max}$ is α/β. With this result in mind, we can now ask the question: Is there an upper limit to human life span?

To explore this proposition, we analyzed data concerning the age of death for all people who died in the state of Rhode Island in 1989 (Gaines and Denny 1993). We began by arranging the data into randomly chosen groups of fifty individuals each and noting the oldest person in each group. These were our extreme values. The mortality data for men ($n = 96$ groups of 50) was considered separately from the data for women ($n = 98$). As is true of average life span, the mean maximal life span of women (98.8 years) was significantly greater than that for men (95.2 years). To remove these differences in mean maximal life span, we expressed each extreme as the deviation from these respective averages.

These deviations were then ranked and used to calculate cumulative probability curves for the maximal age at death for both men and women. The curves were then fitted using eq. (7.30), with the results shown in table 7.1.

TABLE 7.1 A Model for the Maximum Age at Death in Rhode Island

	Males	Females
α	2.854	2.361
β	0.212	0.074
ϵ	−1.267	−1.259

Note: The three parameters given here adjust eq. (7.30) to match empirical data.

In both cases, $\beta > 0$, suggesting that there is indeed an absolute upper limit to human life expectancy. However, the particular value for this bound (α/β) differs strikingly between males and females. The predicted upper limit on maximal age of death for men is 13.5 years beyond the mean, much lower than the suggested maximal age of death for women (31.9 years beyond the mean). Given the mean maximal ages of death in Rhode Island, this suggests that a women could possibly live to be 130 years old, but a man could not live beyond 109. The oldest human ever reliably recorded (as reported in the *Guinness Book of World Records*, Young 1997) was Jeanne Louise Calment, an amazing French woman who was born on February 21, 1875, and died in the summer of 1997 at the very ripe old age of 122. The oldest person in the United States was Carrie White, who died in 1991 at the age of 116. Note that the limits predicted for old age will vary depending on the average maximal age in a given population, and could well vary among populations according to the probability structure of extreme old ages. The oldest man on record was Shigechiyo Izumi of Japan, who died at the age of 120 years 237 days, substantially beyond the age we predict for men in Rhode Island. But Izumi lived in the country with the world's highest average age at death.

Medical and societal changes have dramatically increased the average life span of human beings over recorded history. It would be interesting to compare this analysis of modern life span to comparable analyses of historical data to see if *maximal* life span has changed correspondingly. Current theories of senescence (e.g., Charlesworth 1980; Williams 1957) suggest that while average life span can readily change, maximal life span cannot.

7.7 *Play Ball!*

So far, we have used the statistics of extremes to examine either the physical environment or individual organisms. The same ideas can be used, however, to explore the behavioral interaction among organisms. All sorts of behaviors have random consequences—the interaction of predators and prey, males and females in mating season, competitors for scarce resources—and it would be intriguing to examine the extremes of these processes. Unfortunately, the appropriate data for natural interactions are scarce, and we turn instead to a particular example of human interactions, the game of baseball. In this classic "game of inches," the outcome of a particular contest can indeed depend on chance occurrences. A bunt that appears to be headed foul may hit a pebble and stay fair, allowing a run to score. A gust of wind at the wrong time can turn a home run into a long fly out. Given these sorts of random events, what is the maximum number of games that even a top-notch team can win in a row before chance alone causes them to lose? Conversely, even the worst teams in baseball occasionally get lucky and win a game by chance. What is likely to be the maximum length of a losing streak? Thanks to the penchant of baseball fans for keeping compendious records, the data for an analysis of these questions are readily available.

In this case, the extremes we note are the longest winning and losing streaks in each league, and the interval is a "season."[6] The data are shown in figures 7.11 and 7.12. With one exception, the fluctuations in streak length appear to be both stationary and ergodic, and are thus appropriate for the statistics of extremes. There is a slight (but statistically significant) trend toward shorter winning streaks in the National League, and we have removed this trend before analyzing the data.[7]

Proceeding as before, the longest winning streaks for each league are arranged in ascending order, and the probability of encountering a given length of streak is calculated using eq. (7.29). These cumulative probability functions are then fitted using eq. (7.30) with the results shown in table 7.2. The process is then repeated for losing streaks.

There are several ways in which to explore these results. First, we note that in each case β is substantially less than 0, so that we are dealing with Gumbel Type II cumulative probability curves. Unlike a Gumbel Type III, this type of curve has no well-defined upper bound—the longer the streak, the less likely it will occur, but there is no length of streak for which this likelihood goes to zero.

---

[6] In the period from 1901 to 1990 (from which these data were taken), the length of a full major league baseball season has varied from 154 to 162 games. This small variation in interval size has negligible effect on our conclusions.

[7] Methods used to remove trends from data are discussed in depth in Gaines and Denny (1993).

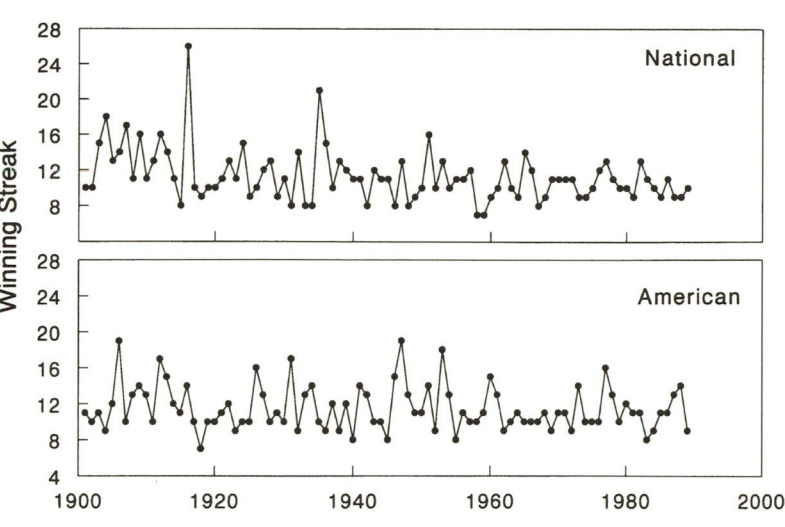

FIG. 7.11 Winning streaks in major-league baseball, 1901–1990. Note the twenty-six-game streak posted by the New York Giants of the National League in 1916. (Data from Wolf 1990.)

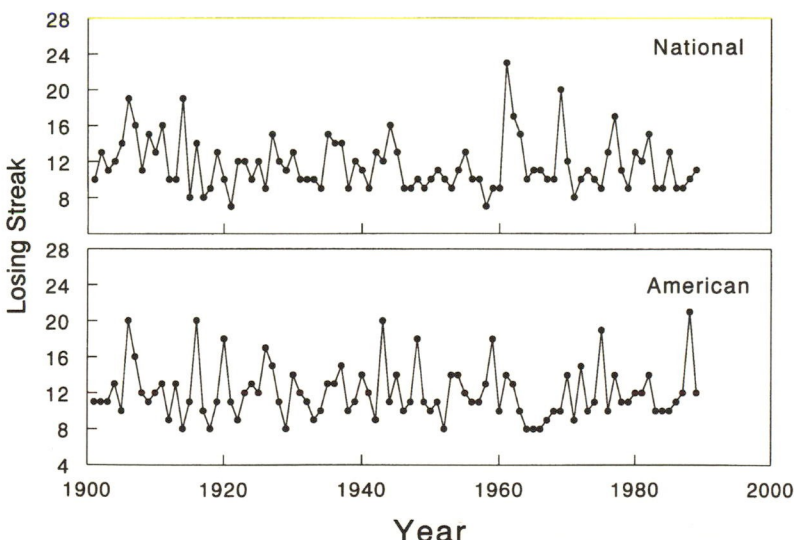

FIG. 7.12 Losing streaks in major league baseball, 1901–1997. The length of streaks is similar between the National and American leagues. (Data from Wolf 1990.)

TABLE 7.2 Extreme Winning and Losing Streaks in Major-League
Baseball

	National League		American League	
	Winning	Losing	Winning	Losing
α	1.963	1.104	1.531	1.318
β	−0.014	−0.093	−0.033	−0.076
ϵ	10.086	10.224	10.332	10.667

Note: The three parameters given here adjust eq. (7.30) to match the history of the sport.

Next we can look at these streaks in terms of their predicted return times, shown in figures 7.13 and 7.14. The two leagues differ substantially in the length of time we would expect to wait between winning streaks of a given length (fig. 7.13). For example, we predict that a winning streak of twenty games will occur in the National League once every 65 years (on average), but that we could expect to wait 119 years between such streaks in the American League. In fact, in the 96-year history of the major leagues there have been two winning streaks of at least twenty games in the National League, the last in 1935, and we are perhaps due for another. There have been no twenty-game winning streaks in the American League, and we wouldn't bet the ranch that one will come along any time soon. In light of these results, the twenty-six-game winning streak posted by the National League's New York Giants in 1916 is truly exceptional, an event we would expect to happen only once in every 429 years. It's worth noting, however, that the Giants finished in third place that year.

The ability to lose consistently seems to be similar between the two leagues (fig. 7.14), and it appears to be easier to maintain a losing streak than a winning streak. For example, in the American League a losing streak of at least twenty games can be expected every 50 years or so, compared to the predicted wait of 119 years between winning streaks of the same length.

Although in this example we have used data from baseball as an illustration of how the statistics of extremes can be applied, the same type of analysis can be applied to biological situations. For instance, it is easy to envision a situation in which predators and their prey encounter each other haphazardly, a lion lying in wait next to a watering hole, for example. If there is some chance involved in whether the predator "wins" a meal (some prey manage to run away before the lion reaches them), we have a system similar in principle to that discussed above. It is possible, for example, that a predator might sustain a long "losing" streak by chance alone. How likely is it that an unlucky predator will starve? The statistics of extremes could tell us.

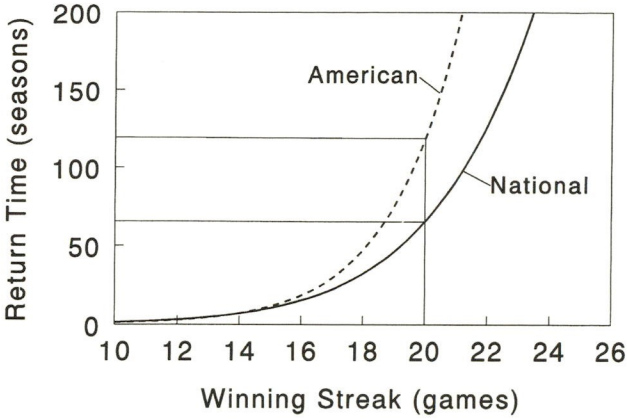

FIG. 7.13 The calculated return time of winning streaks (measured in seasons) for teams in the American and National leagues. For any specified length of winning streak, the return time is higher in the American League.

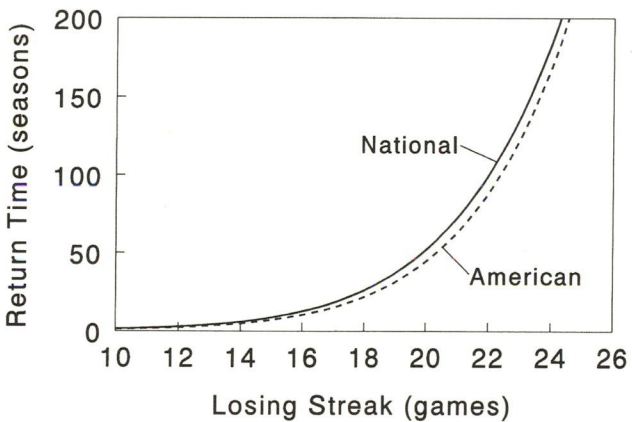

FIG. 7.14 The calculated return time of losing streaks (measured in seasons) for teams in the American and National leagues. Maintaining a losing streak seems to be comparably easy in both leagues.

The interaction between predator and prey provides the excuse for another sojourn into the statistics of baseball. If we view a batter as a predator and the pitcher as his prey, we may ask the question, how likely is it that the predator will win (by getting a hit) a given fraction of his competitive interactions? In particular, how likely is it that we will ever again see a 0.400 hitter? The raw data for this analysis are shown in figure 7.15. The interval is again the "season",

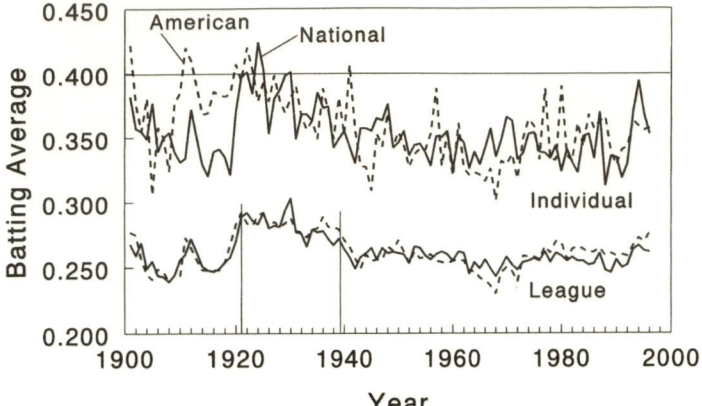

Fig. 7.15 Batting averages (both for the leagues and for individual batting champions), 1901–1997. The league averages in both leagues were exceptionally high between 1921 and 1939. (Data from Wolf 1990 and Thorn et al. 1997.)

and the variable of interest is the maximum individual batting average, where averages above 0.400 are of particular interest. There have been nine 0.400 hitters in the American League and four in the National League since 1901. The last 0.400 hitter was Ted Williams (0.406) in 1941.

In this case, the analysis is complicated by history. Close examination of figure 7.15 reveals a tendency for 0.400 hitters to occur in years when the mean batting average of the league is exceptionally high. This correlation is particularly noticeable in the National League, where for each year in which a batter hit over 0.400 the league average was 1.4 to 3.2 standard deviations above the long-term mean. Furthermore, league batting averages seem not to have varied randomly through time. From 1921 through 1939, the batting averages of both leagues were exceptionally high. How can we analyze the data in light of this history?

First, we can separate the effects of league and individual batting averages. This is accomplished by subtracting the league average from the maximum individual average for each year (fig. 7.16). We can then rank these maximum *deviations*, calculate a cumulative probability function, and proceed as before. The results are given in table 7.3, and they allow us to calculate how likely it is that in a given season the maximum individual batting average will deviate by a given amount from the league average. These probabilities (expressed in terms of the return time) are shown in figure 7.17. The greater variability in individual deviations from the American League average (evident in fig. 7.16)

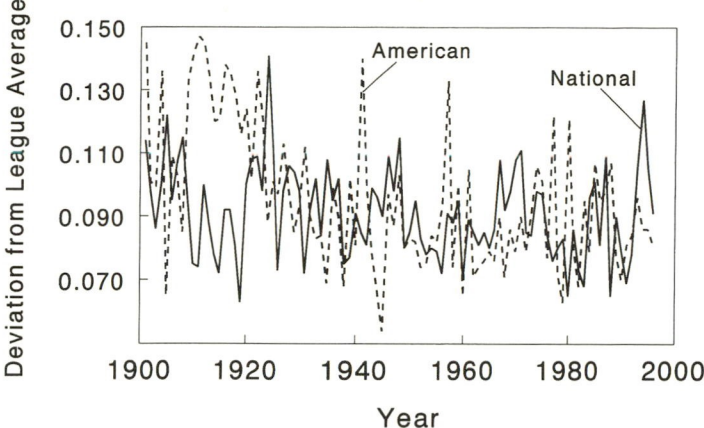

FIG. 7.16 The deviation of maximum individual batting average from the league average, 1901–1997. Larger fluctuations are evident in the record for the American League.

TABLE 7.3 Extreme Deviations from League Batting Average, a Step Toward Predicting When We Will See Another 0.400 Hitter

	National League	American League
α	2.424	1.912
β	0.130	0.015
ϵ	8.516	8.575

Note: The three parameters given here adjust eq. (7.30) to fit the history of the sport.

results in shorter return times for any given level of deviation. It is apparently more difficult to be much better than average in the National League.

But that is only half of the problem. If, for instance, we know the probability that a batter will exceed the league average by 0.100, we still need to know what the probability is that the league average will exceed 0.300. If one of these events occurs, the other must also occur if we are to have a 0.400 hitter. With a little bit of mathematical ingenuity, however, we can arrive at a reasonable answer.

If we exclude the anomalous years 1921–1939 from our data, the league batting averages appear to be stationary through time, allowing us to calculate a reasonable mean and standard deviation of league batting averages for each league. The National League has a mean batting average of 0.256 with a standard deviation of 0.007, while the American League has a mean of 0.257 with a standard deviation of 0.010. If we assume that league averages are normally

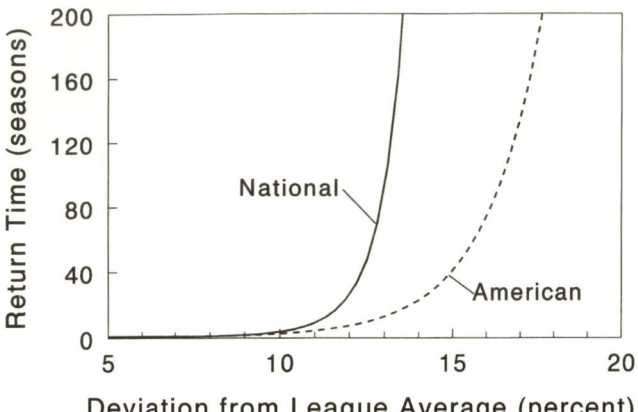

FIG. 7.17 The return time (measured in seasons) for individual deviations from the league batting average. For any specified deviation, the return time is much larger in the National League.

distributed, we can use these parameters to calculate the probability that each league's mean batting average will lie within a given range (between 0.260 and 0.261, for instance). We can then calculate the probability that an individual will exceed the middle of this range sufficiently to hit 0.400. In the example here (league average = 0.2605), a batter would have to hit at a rate 0.1395 above the league average to attain the magic 0.400.

At this point it may help to put these ideas into the jargon of probability theory. Let $P(A)$ be the probability that the league average will have value A. Let $P(B|A)$ be the probability that the maximum individual average will be $B = 0.400 - A$ above the league average. What we want to know is $P(B \cap A)$, the probability for a given level of league performance that an individual's deviation above the league mean will produce a 0.400 hitter. If $P(B \cap A)$ at one level of A is independent from other levels (and it should be), we can add up $P(B \cap A)$ over all A to arrive at our final answer.

We now have all the information we need. The parameters of table 7.3 allow us to calculate $P(B|A)$, and the means and standard deviations of league averages let us calculate $P(A)$ for given ranges of A. Multiplying these values and summing across all A, we come to the following conclusions. In the American league, the probability of a batter hitting 0.400 in a season chosen at random is 0.0438. In other words, we can expect a 0.400 hitter in this league every 22.8 years, or 3.5 such hitters in the 79 years of league play (excluding the glory years 1921–1939). In fact, there have been four 0.400 hitters in the American League outside of the glory years, quite close to our expectation.

The National League is much harder on batters. The probability of batting 0.400 in a season chosen at random is only 0.0023, corresponding to a return time of 433 years! Indeed, outside of the glory years, there have been no 0.400 hitters in the National League. So, if we were the betting sort, we would be willing to place a small wager that the American League will produce a 0.400 hitter sometime soon (a 56-year dry spell for a 22.3-year return time suggests that we are overdue). In contrast, we aren't planning on throwing our money away betting that the National League will produce a 0.400 hitter anytime in the next century. Tony Gwynn (0.394) almost spoiled this prediction in 1994, but close doesn't count!

How can this type of analysis be applied to biology? A pertinent example might be found in the aggressive behavior of sarcastic fringeheads, as described in chapter 2. In addition to wrestling for shelter, these fish square off for mates. Thus, mating success depends on a male's ability to win contests of strength against other males, contests that we assume involve an element of chance. The success of an individual male, and thus the maximal fraction of a population's genes that are contributed by an individual, could be analyzed by the process outlined here. Once upon a time we tried to convince the editors of *Ecology* that this type of analogy between baseball and biology was valid. They were not amused.

We can't leave the subject of baseball without briefly touching on the subject of home runs. Perhaps the most hallowed record in baseball is the number of home runs hit in a single season. The magic number of 60 was set by Babe Ruth in 1927, broken by Roger Maris, who hit 61 in 1961, and shattered by Mark McGwire in 1998 with 70. How unusual was McGwire's performance? To answer this question, we can employ the statistics of extremes.

Beginning in 1921, the number of home runs hit by individual batters in the National League (in which McGwire plays) has gradually inched upward (fig. 7.18A), although there is substantial variation around this tendency. An analysis of the deviations from this trend (shown in fig. 7.18B) allows us to quantify the probability associated with a particular deviation. Based on the trend through 1997, we expected that the National League home run king would hit 45 dingers in 1998. In fact, McGwire hit 70, a deviation of 25. The statistics of extremes estimates that we will encounter a deviation of this magnitude only once every 13,853 seasons! The home run derby of 1998 was truly exceptional.

Exceptional, yes, but not absolute. The curve shown in fig. 7.18B (a Gumbel type III curve) has an α/β ratio of 29.3. McGwire fell four home runs short

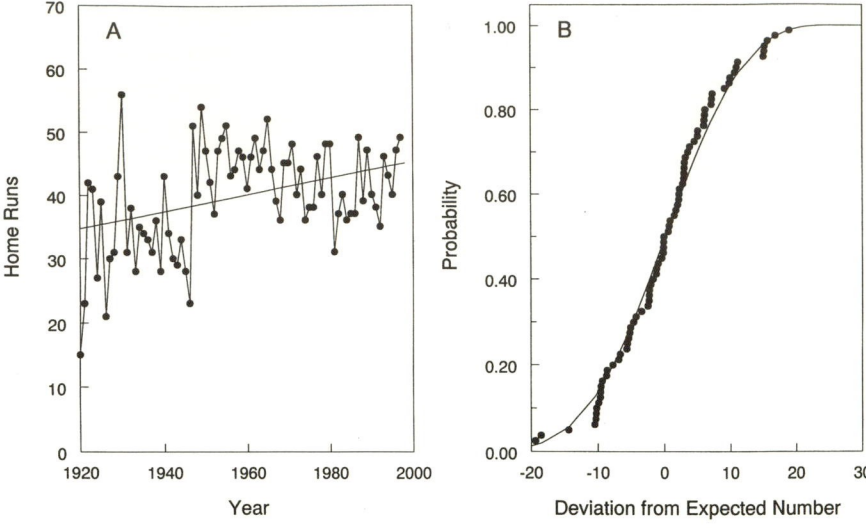

FIG. 7.18 Home runs in the National League. (A) The maximum number of home runs hit in a season by an individual, 1921–1997. The regression line is used to predict the number expected in a given year, and it is the deviation from this predicted value that is used to calculate probabilities. Prior to 1921, baseball strategy did not include the home run, and very few home runs were hit. These early data have not been used in this analysis. (B) The cumulative probability distribution for deviations from the expected number of home runs. The dots are the historical data (1921–1997), and the solid line is the theoretical curve (eq. 7.30) fit to these data. Note that Mark McGwire's 1998 deviation of 25 (not included in this analysis) far exceeds all previous efforts.

of the absolute maximum deviation our analysis predicts, leaving baseball fans something to look forward to.

7.8 *A Note on Extrapolation*

Our analysis above (for instance, our exploration of the likelihood of seeing another 0.400 hitter in our lifetimes) raises an important question. In making the prediction for the National League, the expected return time is 433 years, but we have based this conclusion on only 79 years of data. Is this extended extrapolation something in which we can really place some faith? It seems that we have created more information than we have been given.

In a sense this is true. Because we don't have repeated century-long intervals of major-league play, we don't know for sure that 0.400 hitters will occur in the National League with a mean interval of 433 years. We do, however, have a set of mathematical theories (the statistics of extremes) that tells us how the extreme values of processes must behave. In particular, we know from theory that the extreme values of any process we encounter in nature are likely to asymptote to one of only three distributions. Given this theory, all we have to do is come up with enough information to tell us which of these three distributions we are dealing with, and we can then have reasonable faith in how a given process works. It is this asymptotic behavior of extreme value distributions that allows us to extrapolate reliably well beyond the data in hand.

To give this notion some intuitive "feel," consider the predictions embodied in figure 7.7, the maximal pressure amplitudes expected for Rayleigh-distributed sound waves. If we asked you to extrapolate from the curve given here for the mean maximal value, it would be easy for you to take a pencil and extend the curve to the right by a half inch or so. You might not be exactly correct in your extrapolation, but as long as you didn't draw a line that made a sharp break with the existing smooth curve, you would be close to the true value. But note the scale on the abscissa of this graph. Because time is given in logarithmic units, your half-inch extrapolation has extended the predicted value for the extreme pressure from an interval of a year to an interval well beyond a century! Because the extreme pressure is expected to increase with the square root of the logarithm of time, small extrapolations can extend the predictions from short-term measured data to very long times. The same is true for the three asymptotic distributions derived by Gumbel.

An example of the utility of extrapolation in the statistics of extremes can be drawn from aeronautical engineering. The turbojet and turbofan engines of today's airliners use what amounts to a large multibladed fan to compress air as it enters the engine. Turbulent fluctuations in the air as it is forced into the turbine can interfere with the action of this fan, and occasionally by pure chance the turbulent fluctuations are sufficiently severe to stall the fan and the engine "flames out." This is not a good thing. It's the kind of occurrence that causes the pilot to announce that he has unfortunately "lost an engine" and he will try to restart it, and everyone's knuckles turn white for a few minutes. As you might expect (and certainly hope), engineers have done their best to design turbines that minimize the likelihood of a flame-out. The state of the art is such that engines are rated by the small number of failures expected per 100 hours of flying time.

But therein lies a problem. Say you have just designed a new engine and want to know how likely it is that it will flame out. If you have done a good job, you might have to run the device for hundreds of hours before it fails even once, and you would have to repeat that experiment many times to get a reliable average. This sort of extended experimentation could be expensive as well as inconvenient. There is a quicker way, however. It is well known what extreme turbulent conditions are necessary to stall a turbine, and it is possible to measure the turbulent fluctuations in front of a running engine with great accuracy. So in practice, engineers measure the turbulence levels acting at the fan for just a few minutes, and apply the statistics of extremes to predict the return times of fluctuations severe enough to stall the engine. Even though these predicted return times are measured in hundreds of hours based on mere minutes of data, they turn out to be quite accurate (Jacocks and Kneile 1975). Perhaps the next time you land safely you will have an increased appreciation for the reliability of extrapolations made using the statistics of extremes.

7.9 *Summary*

In chapters 2–5 we explored the utility of probability distributions as applied to individual measurements and asked question such as the following: How far can we expect a molecule to diffuse in a given period? How long will it take a given allele to drift to fixation? In this chapter, we have used the same ideas but have applied them to an examination of the probability distribution of *extreme* values rather than to the underlying population of values from which the extremes are chosen. In cases where the underlying probability density function is known (the Rayleigh distribution, for example), the distribution of extremes can be calculated exactly, allowing us, for example, to speculate with some assurance about the probability that our eardrums will survive the next cocktail party. In many cases, the underlying probability structure of a problem is not known, but the statistics of extremes comes to our rescue. Knowledge that the distribution of extreme values is likely to follow one of only three asymptotic forms allows us to make robust predictions about the return time of rare events.

Examples of the use of the statistics of extremes, as well as many of the fine points of the statistical methods, are described by Gumbel (1958) and Gaines and Denny (1993).

7.10 *Problems*

1. The numbers below are consecutive wave amplitudes measured during a hurricane in the South Pacific. These waves, like most ocean waves, have a Rayleigh distribution of amplitudes, and in this particular case have an average period of 11 seconds. Given the sea surface of which this is a representative example, calculate the following:
  • The probability that the amplitude of a wave chosen at random exceeds 4 meters.
  • The most probable maximum wave amplitude in a sample of 1000 waves.
  • The expected maximal wave amplitude in 24 hours.

0.58	0.69	0.42	0.05	0.61	0.98	1.38	2.14	2.15	0.75
0.24	1.97	2.86	2.48	0.88	0.21	1.10	1.00	0.92	0.80
0.52	0.19	0.88	1.50	0.90	0.72	1.35	0.82	0.89	0.55
0.79	0.39	0.81	1.31	0.99	1.56	2.05	1.40	0.80	1.16
1.19	1.00	1.34	0.73	0.50	1.08	0.76	1.31	1.01	0.75
1.37	1.46	1.30	0.87	0.36	0.90	1.09	0.27	0.54	1.59
1.36	0.70	1.48	1.87	1.23	0.68	0.75	0.51	0.33	0.36
0.63	1.45	1.21	2.27	3.15	2.46	3.09	2.06	0.50	0.99
0.72	0.74	1.17	0.58	0.63	0.71	0.94	1.27	0.65	0.24
1.21	1.82	1.31	1.77	2.74	2.76	1.37	1.40	1.70	1.23

2. The numbers below are maximum wind speeds. Each entry is the maximum speed recorded in a 2-hour interval.

- Calculate and graph the cumulative probability function for these maxima.
- We have fitted these data with eq. (7.30), with the following results:

$$\alpha = 1.4306$$

$$\beta = 0.005279$$

$$\varepsilon = 2.8307.$$

Using these parameters, calculate the return time (in hours) for a wind of 10 m s^{-1}. Discuss why you should (or should not) be confident of this prediction.

3.696	2.355	2.478	0.741	2.638	5.763	5.078	3.213	2.920	5.253
2.490	5.005	2.667	3.104	5.062	6.303	4.591	2.581	2.011	6.226
5.995	2.721	4.598	3.042	2.569	4.076	1.328	2.242	2.358	3.886

- What is the absolute maximum wind speed predicted by this analysis?

8

Noise and Perception

Biologists are accustomed to dealing with random fluctuations. The size of a population may fluctuate, the physical environment varies through time, and (being complex creatures) both plants and animals exhibit behaviors that can appear to be stochastic. In many of these cases, it is possible to cling to the hope that if we knew just a little bit more about how the population, environment, or organism worked, we could predict the fluctuations that now appear random. There are aspects of life, however, where random behavior, what we will call *noise*, is unavoidable. In this chapter we will expand our repertoire of stochastic processes by examining noise and how the presence of noise can place fundamental limits on organisms' ability to sense their environment. Along the way we will explore the interaction between the detail that your eyes can detect and how fast they can detect it, discover a reason why small lizards and frogs have eardrums almost as large as those of humans, and attempt to determine the limit to how small nerve cells can be before they can no longer function reliably.

8.1 *Noise Is Inevitable*

Statistical mechanics is the branch of physics that studies the random motion of particles. It is a highly evolved science, with a long list of successes (including the kinetic theory of gases). One of the most important results of statistical mechanics is that there is a lower limit to the average energy of an object, a limit set by the absolute temperature (Reif 1965). This constraint is true for all forms of mechanical energy, with the following result. For any manner in which an object can move or deform,

$$\overline{\text{energy}} = \frac{kT}{2},\tag{8.1}$$

where k is once again Boltzmann's constant, 1.38×10^{-23} joules per kelvin. We have already encountered an example of this principle in chapter 5 when we noted that the average kinetic energy of a molecule in a gas or liquid was set by the temperature,

$$\frac{\overline{mu_x^2}}{2} = \frac{kT}{2}. \tag{8.2}$$

Here, m is the mass of the molecule and u_x is its speed along the x-axis. Analogous equations hold for motion along the y- and z-axes. Given that the mass of a molecule is constant, we can conclude that

$$\overline{u_x^2} = \frac{kT}{m}$$

$$\sqrt{\overline{u_x^2}} = \sqrt{\frac{kT}{m}}. \tag{8.3}$$

Thus, the root mean square velocity is proportional to the square root of the ratio of temperature to mass. For the relatively high absolute temperatures found on Earth and the small mass of molecules, this relationship requires that the molecules of a gas or liquid move with velocities measured in the hundreds of meters per second. The inevitable result of molecules zipping about at such speeds is that these small particles collide with each other many times per second, a process that makes it impossible to predict the precise path that any individual molecule will take. Thermal agitation (or thermal "noise") thus leads inevitably to the random motion of molecules that held our attention in chapters 5 and 6.

A similar relationship holds for objects that cannot move freely but rather are tethered in place by springs. For example, in our consideration of rubber elasticity in chapter 6, we showed that the potential energy of a linear, Hookean spring is

$$\text{spring potential energy} = \frac{\mathcal{K}x^2}{2}, \tag{8.4}$$

where $\mathcal{K}$ is the stiffness of the spring (N m^{-1}) and x is its displacement from its resting position. The relationship of eq. (8.1) thus tells us that

$$\frac{\overline{\mathcal{K}X^2}}{2} = \frac{kT}{2}, \tag{8.5}$$

from which we can calculate that

$$\sqrt{\overline{X^2}} = \sqrt{\frac{kT}{\mathcal{K}}}. \tag{8.6}$$

The root mean square displacement is proportional to the square root of the ratio of temperature to stiffness.

Here we have calculated the average displacement on the tacit assumption that x varies through time. But could x be constant? If one were to consider eq. (8.6) alone, one might be tempted to conclude that a spring could fulfill its energy requirement with a constant displacement $x = \sqrt{kT/\mathcal{K}}$. But the fact that energy can be stored in a spring does not remove the requirement that an object's kinetic energy (the energy of its mass in *motion*) must also be in accordance with the temperature. In other words, if the spring is attached to a mass (or has mass itself), it must bounce around in a fashion such that its average kinetic energy is equal to $kT/2$ while at the same time its mean spring potential energy is *also* equal to $kT/2$. For a spring-mass system, this can be accomplished by trading kinetic and potential energy such that when the spring is maximally stretched, the mass's velocity is 0, and when the mass's velocity is maximal, the spring is not stretched. For example, a molecule in a solid is held in place by the springlike hold of the chemical bonds that attach it to its neighbors. But at any temperature above absolute zero it cannot remain stationary,[1] and it rattles around in three dimensions, alternately having a high kinetic energy and a high spring potential energy. The average potential energy stored in the deflection of the molecule's chemical bonds is set by the bonds' stiffness and the temperature. The lower the stiffness of the bonds, the larger the average deflection of the molecule. Similarly, the average kinetic energy due to the mass's motion is set by the mass and the temperature; the smaller the mass, the higher the average speed.

The basic relationship between energy and temperature given by eq. (8.1) does not apply solely to molecules. For example, when the small spores of fungi are suspended in water and viewed under a microscope, they are seen to move about in a jerky, random fashion. This "Brownian" motion is a larger-scale version of the motion undergone by gas molecules (Einstein 1905). Another example concerns the thermal fluctuation of sensory cilia—for instance, the cilia in your inner ears that are used to sense sound. These stiff, rodlike structures are supported by a basal membrane that acts as a spring. To a first approximation, the torque required to deflect a cilium is a linear function of the angle to which the cilium and its supporting membrane are bent. Thus, by analogy to a linear spring, the spring potential energy stored by the angular deflection of a cilium is

$$\text{spring potential energy} = \frac{\mathcal{S}\theta^2}{2}, \tag{8.7}$$

---

[1] Actually, molecules would move even at absolute zero due to the Heisenberg uncertainity principle, a quantum effect that prohibits our knowing both the precise location and speed of a molecule.

where S is the torsional stiffness of the structure (newton meters per radian) and θ is the angular deflection. eq. (8.1) thus tells us that

$$\frac{\overline{S\theta^2}}{2} = \frac{kT}{2},$$

(8.8)

and

$$\sqrt{\overline{\theta^2}} = \sqrt{\frac{kT}{S}}.$$

(8.9)

In other words, the cilium can be expected to have some average fluctuation in its angular position set by the temperature. The lower the torsional stiffness of the cilium, the larger its average fluctuation. As we will see later in this chapter, this thermal noise may affect our ability to hear very soft sounds.

Thermal agitation in all its many forms is thus revealed as an inevitable source of noise with which organisms must cope. It is, however, not the only such source. Beginning with Max Planck in 1900, physicists have uncovered a bewildering set of behaviors that are most evident at very small spatial scales. This is the realm of quantum mechanics noted in chapter 1, one particular result of which is relevant here.

A light beam can be thought of as a stream of particles in which each particle, or *photon,* has a fixed amount of energy. If an object (a lightbulb, for instance) produces light, it is quite reasonable to measure the rate at which energy is being radiated by counting the average number of photons produced per second. But even when this average is constant, the precise timing of the release of individual photons is unpredictable. For any object emitting light, the interval between the emission of individual photons varies randomly. We seldom notice this effect because in typical daytime lighting conditions, billions of photons reach our eyes each second. Given this large sample, the fluctuations are infinitesimal relative to the mean rate at which photons are delivered to our eyes, and the light appears constant. If, however, a light beam is very dim so that only a few photons reach our eyes per second, the inevitable second-to-second variation in the arrival of photons (one example of an effect known as *shot noise;* Rice 1944, 1945) can become apparent.

The inevitability of random fluctuations at small scales (the scales of photons to cilia) means that when organisms try to sense the smallest details about their environment they must do so in the presence of "noise." Knowing what we do about the physics of noise, what can we learn about how biology has evolved to cope with it? What are the physical limits to the detail of sensation, and how close to these limits have organisms come? In the rest of this chapter, we will cover four heuristic examples of the interaction of noise and the senses:

1. The limits to sight in dim light.

2. The effects of thermal noise on the eardrum.

3. The effects of channel noise in nerve cells.

4. A process by which noise can actually enhance our ability to sense the world.

We first explore the problem of how the random arrival of photons affects sight.

8.2 *Dim Lights and Fuzzy Images*

The basic elements of a vertebrate eye are shown in figure 8.1. Light from an object is focused by the lens onto a retina composed of receptor cells (either rods or cones). We are concerned here with dim light, to which rods are most sensitive, so the cones will take no part in this discussion. Each rod cell detects whether light is falling on it, and if light is detected, a nervous signal is sent off to adjacent rods as well as toward the brain. This output from each cell defines the primary "nervous image" perceived by the eye. This image is considerably massaged by the nervous system, and much that is important to the process of sight occurs after the primary image is recorded (for example, see the review by Baylor 1995). For present purposes, however, we are only concerned with the process up to the point at which a rod cell sends off a signal. We will simplify matters even further by assuming that each rod can be in one of only two states. It can be "on," indicating that it has recently detected the arrival of a photon, or it can be "off," indicating that no photon has arrived. Given this assumption, we can think of the primary output from the retina as a high-contrast, black-and-white image of the object on which the eye is focused—white for those regions of the image where light has been detected, black for those regions from which no light has arrived. So far, so good.

The problem is that light is not continuous—as we have noted, it is composed of individual photons. Therefore, the light originating at some specific point on

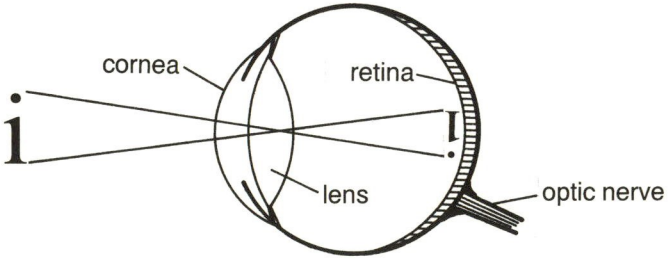

Fig. 8.1 A schematic diagram of the vertebrate eye. The lens projects the image of an external object onto the retina.

the external object and falling on a specific rod comprises a stream of photons whose arrival times we cannot predict. Therein lies a potential problem.

Vertebrate eyes are *extremely* sensitive to light. For example, in experiments conducted in the 1940s, humans were placed in a dark room and asked to respond when they saw an experimental flash of light. By flashing ever dimmer lights, the experimenters found that their subjects could detect the arrival of 2–8 individual photons. Since it was unlikely that all of these photons fell on a single rod cell, Hecht et al. (1942) and van der Velden (1944) concluded that rod cells can respond to the absorption of *single* photons.[2] This suggestion was proven some years later by Baylor et al. (1979) working with a preparation of isolated frog rod cells.

The ability of rods to respond to single photons leads to the following scenario. Imagine that you are trying to form an image of the world around you in a very dim light (starlight, for instance, or the light of a crescent moon on an overcast night). If a photon arrives at a rod cell in your retina, you know that light is being emitted from that part of the image. If, however, no photon has arrived at a particular rod, you have a decision to make. Is that part of the image really black? That is, will you never receive a photon from that location? Or are you just experiencing the natural interval between photon emissions, waiting for the photon to arrive? How can you (or your nervous system) appropriately decide?

8.3 *The Poisson Distribution*

To explore this question we must turn to statistics, and we begin by returning to our old workhorse, the binomial probability distribution. Recall from chapter 3 that

$$P(i \mid n, p) = \frac{n!}{i!(n-i)!} p^i q^{n-i}, \tag{8.10}$$

where n is the total number of trials, p is the probability that a "success" will occur in a given trial, and i is the number of successful trials. As always, $q = 1 - p$. We have seen that when n is large and $p = q = 0.5$, the binomial distribution "asymptotes" to the normal distribution. There are other asymptotic forms of the binomial distribution, however. Consider, for instance, a situation in which the probability of a "success" is small (that is, $p \ll 1$) , so that the number of trials in which a success occurs (i) is much less than the total number (n). In this case, when n is large,

$$P(i \mid \mu) = \frac{\mu^i}{i!} \exp(-\mu). \tag{8.11}$$

[2] For a thorough and readable review of these findings, see Cornsweet (1970).

Here μ is again the mean of the distribution. This is the *Poisson distribution*, a brief derivation of which can be found in the appendix to this chapter.[3] Note that this is a probability distribution (as opposed to a probability density); i here must be an integer greater than or equal to zero.

The Poisson distribution is of immediate use to us. If the light is dim, the probability is small that any given rod cell in your retina will be hit by a photon in some small interval of time. Thus, if we define the detection of a photon as a "success," p is small. As a result, it is appropriate to use the Poisson distribution to calculate the chance of making an error in deciding whether some spot in our visual image is really dark.

Look at the problem this way. Suppose that through repeated long-term measurements we know μ, the average number of photons that will reach a specific rod cell in a given period if that cell is indeed being exposed to light from the external object. We now expose that cell to a certain part of the image and are asked to decide whether or not there is light coming from that location. After waiting a standard period (a second, say), no photon has arrived, and we decide that portion of the image is indeed dark. What is our chance of being mistaken? In essence, we are asking what the probability is of having no successes (the arrival of no photons) when the mean is actually nonzero (in this case, μ). This is a question that can be answered immediately using the Poisson distribution:

$$P(0|\mu) = \frac{\mu^0}{0!} \exp(-\mu). \tag{8.12}$$

Recalling that $0! = 1$, we see that

$$P(0|\mu) = \exp(-\mu). \tag{8.13}$$

In other words, the larger the mean number of photons absorbed in a standard interval, the lower the probability of not having any photons arrive, and the lower the chance of making an error in forming our image.

We can take this analysis one step further. The mean number of photons absorbed by a region of the retina is set by four factors:

$$\text{mean number of photons absorbed, } \mu = \mathcal{I} A \alpha t, \tag{8.14}$$

where $\mathcal{I}$ is the *irradiance* (photons per time per area), A is the area of the retina exposed to light, α is the *quantum yield* (the probability that a photon hitting the cell will be absorbed), and t is time. The finest detail in the image can be resolved when the area over which we count photons is as small as possible.

[3] The Poisson distribution is named for Siméon-Denis Poisson (1781–1840), a French mathematician who made major contributions in celestial mechanics, electromagnetic theory, and probability.

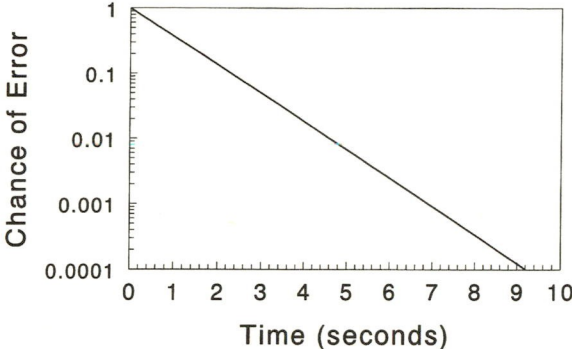

FIG. 8.2 The chance of making a visual error (black instead of white) decreases exponentially with the time you view a scene (eq. 8.16). The calculation represented here assumes that on average one photon per second is absorbed by a rod cell.

Therefore, if we are interested in a detailed assessment of the image, we set the area in this expression to that of a single rod cell. Expressing the area of a cell in terms of its radius, we can rewrite eq. (8.14) as

$$\mu = \mathcal{I}\pi r^2 \alpha t. \tag{8.15}$$

In vertebrates, rod cells have a radius of about 0.38 μm.

We can now write an expression that lets us calculate the chance of mistakenly thinking that a portion of the image is black when it is really white:

$$\text{chance of error} = \exp\left(-\mathcal{I}\pi r^2 \alpha t\right). \tag{8.16}$$

In other words,

- the higher the irradiance,
- the greater the quantum yield,
- the bigger the rod cell,
- and the greater the time you look at the image,

the smaller the chance of making an error. For example, when $\mathcal{I}\pi r^2 \alpha$ is one photon per second, the chance of error as a function of time is shown in figure 8.2. In one second, there is an $e^{-1} = 37\%$ chance of making an error. In five seconds, this chance is reduced to less than 1%. In other words, at this low level of irradiance, you can accurately see the finest detail only if you look at an image for several seconds. At shorter times, you will inevitably make errors in interpreting the image, and the image is therefore likely to look "fuzzy."

This effect is simulated in figure 8.3. In making this figure, we began by creating a complete binary image of a small fish swimming through a coral reef (essentially the image at the lower right-hand corner of the figure). The image was divided into discrete small blocks known as pixels, and we taught our computer to deal with these pixels separately. By manipulating the value of the pixels, we could then simulate the subset of the information from the complete image that would be available after a given amount of time. To produce each of the incomplete images shown here, we selected a random number between 0 and 1 for each pixel. If the number was less than the chance that a photon would have arrived from that portion of the image in the allotted time, the pixel was set to be black. If the random number was greater than the chance that a photon would have arrived *and* the pixel was white in the complete image, the pixel was set to be white. As you can see, the simulated image from short times is quite fuzzy; it is difficult to pick out the details. As time progresses, the image becomes clearer.

There is an obvious way to reduce the time required to make an accurate decision about the presence or absence of light in a certain portion of the image. By increasing the area over which you search for photons (that is, by increasing r in eq. 8.16), you can reduce the chance of making an error. For example, you could "gang" together several rod cells into one nervous unit, thereby increasing the area over which light is absorbed. Unfortunately, by ganging rod cells, you also reduce the resolution of the image. There is thus an intrinsic trade-off between the time required to sense a detailed image and the amount of detail that can be sensed. Rod cells in humans do indeed seem to be ganged to a certain extent, but with an adverse effect on our ability to judge motion at low levels of light (Gegenfurtner et al. 1999).

There is a less obvious (but more desirable) way than ganging cells to decrease the chance of making a visual error. If the effective quantum yield α can be increased, $\mathcal{I}\pi r^2 \alpha t$ is increased, and the chance of error is reduced. The quantum yield can be increased by packing more absorbing molecules into each cell, and (as we will see later in this chapter), rod cells are indeed chock full of absorbing molecules.

Another trick has evolved in a variety of animals. This stratagem has been observed by anyone who has walked through the woods at night with a flashlight or driven down a dark country road and seen the ghostly bright reflection from the eyes of deer, raccoons, and cats. These and many other nocturnal animals have a *tapetum*, a mirrorlike layer of cells, behind their retina. Now, if you place a mirror behind a rod cell, any photon that passes through the cell without being absorbed can be reflected back into the cell. This reflection gives the cell a second chance to absorb the photon, and the quantum yield is thus substantially enhanced. The tapetum is therefore an effective means of decreasing the time

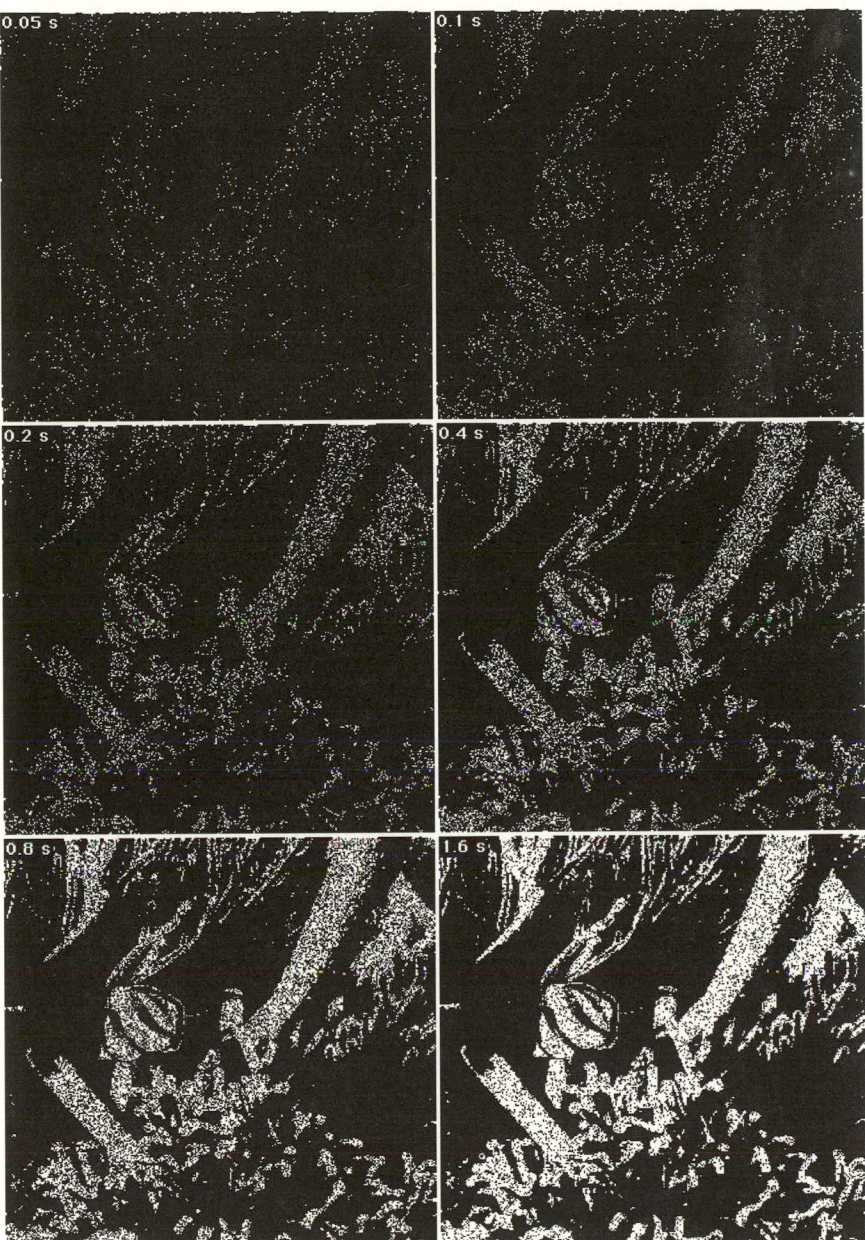

FIG. 8.3 Scene from a coral reef: a simulation of how visual acuity increases through time. Each panel presents the information available to the eye after a specified interval. Initially, the image is very vague, but after about 0.8 seconds the details of the scene are evident. Note, however, that in 0.8 seconds a real fish would have moved a considerable distance.

it takes an animal to make accurate decisions about the fine details of a visual image.

Rod cells can also increase the probability of absorbing a photon by changing their shape. In dim light, rods actively elongate, thereby increasing the path length of photons as they pass through the retina (see Ali and Klyne 1985). The change in length requires time, and this is one reason that it takes a while for our eyes to become "dark adapted."

Note that there is a lower limit to the response time of rod cells, set by the rate at which chemical reactions in the cell can effectively record the arrival of a photon. For rods of primates (of which we are one) this response time is approximately 0.3 seconds (Schnapf and Baylor 1987). This is one reason it becomes increasingly difficult to catch a baseball or hit a pitch near dusk. Not only does the image of the ball get fuzzier as the light gets dimmer, but by the time your rod cells have recorded the image of the ball, the ball has moved a considerable distance. For example, a major-league fastball traveling at 90 miles per hour (41 m s^{-1}) moves 12.3 meters in the time it takes a rod to respond.[4]

8.4 *Bayes' Formula and the Design of Rods*

So far we have assumed that rod cells are unerring in their detection of photons—if a photon has not been absorbed by the cell, no signal is sent to the nervous system. We have also noted, however, that rods can detect single photons. By "detect" we mean that the molecular machinery of the rod cell is capable of amplifying the minuscule energy of the single absorbed photon and transducing this signal into a vastly more powerful nervous impulse for transmission to the brain. As with any amplifier "jacked to the stops," we might expect that there is some noise associated with the detection of photons. What are the consequences if a rod cell makes a mistake every once in a while and reports the arrival of a photon when in fact none has arrived?

To explore this question, we return to the calculations we made in chapter 2 when discussing Bayes' formula and the detection of a rare disease. At that time (eq. 2.9), we saw that Bayes' formula for an "either/or" situation is

$$P(X|Y) = \frac{P(Y|X)P(X)}{P(Y|X)P(X) + P(Y|X^c)P(X^c)}. \tag{8.17}$$

Recall that X^c is the complement of X. Restating this equation in terms of the problem at hand, we see that the probability that a photon has actually struck

---

[4] Cone cells respond considerably faster, so this problem is minimized under the daylight conditions in which cones are the primary source of visual information.

the cell (Ph) given that a nervous signal (S) has been produced is

$$P(Ph|S) = \frac{P(S|Ph)P(Ph)}{P(S|Ph)P(Ph) + P(S|Ph^c)P(Ph^c)}. \tag{8.18}$$

For a rod cell, the nervous signal consists of the release of a neurotransmitter to a ganglion cell, a response that we will discuss in greater detail below. Now, the probability that a nervous signal is produced if a photon strikes the cell, $P(S|Ph)$, is governed largely by α, the quantum yield of the rod cell. For the sake of argument (and to keep the math simple), let's assume that this probability is 1; that is, let's assume that if a photon arrives at a rod cell, the rod invariably causes a signal to be sent toward the brain. The actual quantum yield is about 0.6 (Baylor et al. 1979).

We are interested here in situations where light is dim and decisions must be made about individual photons. Therefore, we assume that the probability that a photon arrives in a specified interval, $P(Ph)$, is very small. Now, if $P(Ph)$ is very small, $P(Ph^c) \cong 1$, and for simplicity we will assume that it is exactly 1. Given these assumptions, eq. (8.18) can be simplified and rearranged:

$$P(S|Ph^c) \cong \frac{P(Ph)[1 - P(Ph|S)]}{P(Ph|S)}. \tag{8.19}$$

In other words, the allowable chance of a rod making an error (by producing a nervous signal when no photon has struck the cell) is set by the probability that a photon arrives, $P(Ph)$, and the level of "certainty" that we require of the system, $P(Ph|S)$. For example, if we need to be 99% certain that when a signal is produced, a photon has actually arrived:

$$P(S|Ph^c) \cong \frac{P(Ph)(1 - 0.99)}{0.99} = 0.01 P(Ph). \tag{8.20}$$

If we are to see accurately when the illumination is very dim, for instance when the probability of a photon arriving in any given second is 0.1, the probability that a signal is produced in the absence of a photon must be less than 0.001. Thus, the rod cell can accidentally produce a signal only once in every 1000 s. The lower the light level $(P(Ph))$ and the higher the required accuracy $(P(Ph|S))$, the smaller the allowable error in the visual system.

8.5 Designing Error-Free Rods

How has this extraordinary level of accuracy been achieved in nature? The story is a bit complex, but interesting, and it has implications beyond just the response of rod cells. In a nutshell, here is how it works (Schnapf and Baylor 1987; Stryer 1987).

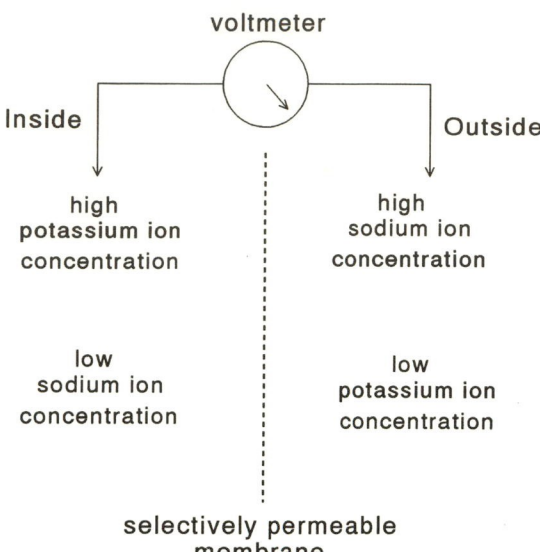

FIG. 8.4 The ionic properties of an animal cell. A selectively permeable membrane separates solutions with different ionic compositions.

8.5.1 THE ORIGIN OF MEMBRANE POTENTIALS

As with many cells in the nervous system, the response of a rod cell to an external signal is intimately connected to the voltage difference across the cell membrane. It is possible (and intriguing) to account for the membrane potential by working from first principles (see, for instance, Hille 1992). That story is long and somewhat technical, however, and for our purposes it is sufficient to pick up the tale in the middle.

Consider the situation shown in figure 8.4. The inside of a cell is separated from the outside by a selectively permeable membrane. The selectivity here is governed by the presence of ion channels. Each channel is an integral membrane protein with a water-filled pore in its middle that allows specific inorganic ions to cross from one side of the membrane to the other. For example, potassium channels are structured such that potassium ions can readily pass through, but other ions are effectively blocked. Furthermore, the shape of channels can change, affecting the channels' permeability. Channels can either be "open" (allowing ions to cross the membrane) or "closed" (preventing molecules from passing). For the time being, let's assume that all the membrane ion channels are closed.

Here we are concerned with the electrical potential difference between the two sides of the membrane. This potential can be measured using a sensitive voltmeter with one probe outside the cell and the other in, as shown in figure 8.4.

To begin, we assume that the solutions both inside and outside the cell are electrically neutral (that is, they each have equal numbers of positive and negative charges).

Now let's consider ionic concentrations, in particular the concentrations of potassium ions (K^+) and sodium ions (Na^+). In living organisms, the concentrations of these important ions are different inside the cell than outside. For reasons that we will not explore here, in animal cells the potassium ion concentration inside the cell is much higher than outside, while the sodium ion concentration is much higher outside than in. It is the interaction of these concentration gradients with the selectivity of the cell membrane that produces the membrane potential.

First, let's consider the potassium gradient. Because the K^+ concentration is higher inside the cell than outside, there is a tendency for potassium to diffuse down this gradient and out of the cell. If all the potassium channels are closed, this can't happen, but if we allow even a few potassium channels to open, potassium ions flow outward. This diffusive flow can go only so far, however. Each potassium ion is positively charged, so the outward flow of K^+ through the open channels amounts to an outward flow of positive charges. Given that the inside of the cell had equal numbers of positive and negative charges before potassium started to flow, the flow of positive charges out of the cell leaves a net surplus of negative charges in the cytoplasm. This net negative electrostatic charge tends to impede the escape of positive ions from the cell, and an equilibrium is reached when the diffusive potential for K^+ ions to move out of the cell is just matched by an electrical potential for K^+ ions to be drawn back in. For the potassium concentrations typical of animal cells (about ten times that of the external medium), this equilibrium is reached when the inside of the cell is about 60 millivolts (mV) negative relative to the outside. We will refer to this voltage as the potassium potential, V_K. In summary, when the cell membrane is permeable to potassium ions (and *only* potassium ions), the membrane potential is −60 mV.

Let's now return to our starting point as shown in figure 8.4. All channels in the membrane are closed, the potassium concentration is high inside the cell, the sodium concentration is high outside, and both the cytoplasm and external fluid are electrically neutral. What happens if we now open a few sodium channels while leaving the potassium channels closed? The result is analogous to the one we just described for potassium ions, but in the opposite direction. Because the Na^+ concentration is high outside the cell, there is a potential for Na^+ to flow in, and the resulting flow of charges leads to a net *positive* charge within the cell. Again, the diffusive flow of ions is resisted by the separation of electrical charges, and an equilibrium is reached when the electrical and diffusive potentials are equal and opposite. For the sodium concentration differences normally found across cell membranes, this sodium potential, V_{Na}, is about +60 mV. In other words, if only sodium channels were open, the inside of the cell would be about 60 mV positive relative to the outside.

In real cells, the inward flux of sodium through channels in the membrane would eventually lead to a sufficient accumulation of sodium in the cell to substantially reduce V_{Na}. As a means to avoid this "wind down," cells have proteins that actively pump sodium out of the cell. These sodium/potassium ATPases use the chemical energy stored in ATP to exchange internal sodium ions for external potassium ions.

Given the counteracting potentials due to potassium and sodium, what is the actual potential of the cell membrane? It depends on the relative permeability of the membrane to potassium and sodium ions. For instance, if many more potassium channels than sodium channels are open, there will be much less resistance to the flow of potassium out of the cell than to the flow of sodium in. As a result, the membrane potential will be set primarily by V_K; it will be substantially negative. Just the opposite is true if there are many more sodium channels open than potassium channels; the cell membrane potential will be set primarily by V_{Na}; it will be less negative and may even be positive. The membrane potential can also be affected by the flow of other ions, although the magnitude of their effect is generally small.

8.5.2 MEMBRANE POTENTIAL IN ROD CELLS

We now apply this information to the membrane potential of rod cells. Each rod cell has a large population of sodium channels, and it is the status of these channels that is the primary determinant of membrane potential.[5] Each channel flickers open and closed under the influence of thermal agitation, but the probability that any individual channel is open is affected by a variety of factors. Here we are primarily concerned with the intracellular concentration of a substance called cyclic guanosine monophosphate (cGMP). In the dark, a rod cell has a relatively high level of cGMP, and in the presence of this high concentration, sodium channels in the cell's membrane have a high probability of being in their open configuration. Because the probability of any one channel being closed is low and there are many channels present, the probability of all channels (or even a large fraction of channels) being closed is very small. Thus, in the dark (because most of its sodium channels are open) the voltage difference across the rod cell membrane is small.

---

[5] Calcium also flows into the cell, but to keep the explanation simple, we will refer here only to the flow of sodium.

Now, the rod contains a compound known as 11-*cis* retinal, a small pigment molecule that is attached to a protein called opsin. Under the appropriate circumstances, as we will see, opsin can act as an enzyme. The combination of pigment and protein is called *rhodopsin* and it inhabits an internal membrane in the rod.

A photon arrives. If the photon is absorbed by rhodopsin, the retinal is transformed from its *cis* to its *trans* configuration. In other words, the absorption of a photon causes rhodopsin to change its shape. There are approximately 10 million rhodopsin molecules in each rod cell, so the probability of capturing a photon as it passes through the cell is relatively high (about 0.6, as we have noted).

The change in shape of the rhodopsin (its photoisomerization) starts a cascade of events. First, the newly transformed rhodopsin molecule (now an active enzyme) catalyzes the activation of a G-protein known as *transducin*. This activation involves the splitting of the transducin from its inactive form (a trimer composed of α, β, and γ subunits) into two—a G_α monomer and a $G_{\beta\gamma}$ dimer. The G_α molecules produced by this fission each associate with a molecule of guanosine triphosphate (GTP) to form the active G-protein. Because the isomerized rhodopsin acts as a catalyst in this process, many active G-proteins can be produced as the result of a single photon absorption. Typically, four hundred active G_α's are created per photon absorbed.

The cascade then continues, with two effects. First, the activated G_α molecules act as a GTPase. In other words, the activated G-proteins act to turn themselves off by breaking down the GTP molecules with which they must associate to be active. As a result, the absorption of a photon can cause only a very brief pulse of high active-G_α concentration. Second, while they themselves are activated, G-proteins activate a phosphodiesterase. As a consequence, during the brief pulse of G-protein activation, the activated phosphodiesterase is in relatively high concentration, and it has an important effect—it catalyzes the rapid breakdown of cGMP. The net result of all this activity is that for a short period after a photon is absorbed, the concentration of cGMP in the rod cell is drastically reduced.

Now, it was the presence of high levels of cGMP that initially elevated the probability that sodium channels would be in their open configuration. When the concentration of cGMP decreases after a photon is absorbed, the probability is decreased that sodium channels are open, and the amount of sodium flowing into the cells through these channels is consequently reduced. The reduction in sodium influx (inward current) results in an increase in the voltage difference across the cell membrane. The cell membrane becomes hyperpolarized.

This change in membrane potential in turn causes the rod cell to reduce the rate at which it releases neurotransmitter molecules to its neighboring bipolar nerve cell. The reduction in neurotransmitter at one end of the bipolar cell causes the increase in release of neurotransmitter at the other end of the cell, and this increased release of neurotransmitter influences the ganglion cell to which the bipolar cell is attached. The form of this influence depends on the circumstances. In some cases, the release of neurotransmitter will tend to cause the ganglion cell to increase the rate at which it fires action potentials. In other cases, the effect will be to decrease the rate of firing. In all cases, however, information has been provided to the central nervous system.

At this point the cascade has run its course. The reception of a photon has resulted in communication with the central nervous system. The rod cell has then turned itself back off (that is, the cGMP levels are again elevated and the membrane potential is low) and the rod is ready to detect another photon. As we noted earlier, this whole process takes about 0.3 seconds.

There are two aspects of this amplification process that ensure that it is highly unlikely that a mistake will be made. First, the rod cell works backwards from most sensory cells. Rather than being in a quiescent mode (with sodium channels closed), a rod in the dark is in an active state, with most of its sodium channels open. In effect, this increases the size of the sample the cell looks at when deciding whether a photon has arrived, thereby reducing the chance of error. For example, consider a cell with n channels, each with a probability p of being open at any given time. From our knowledge of binomial experiments (of which this is one), we can calculate μ (the mean number of channels open, a value proportional to the membrane voltage) and σ (the standard deviation of the number of open channels):

$$\mu = np$$
$$\sigma = \sqrt{npq}. \tag{8.21}$$

At any instant, the membrane voltage is responsive to the state of all channels, so the effective sample size is n. Thus, the standard error of the mean number of channels open is

$$\text{standard error} = \frac{\sqrt{npq}}{\sqrt{n}} = \sqrt{pq}. \tag{8.22}$$

This is a measure of the expected fluctuation through time in the number of open channels. The effect that these fluctuations will have, however, depends on the level of the fluctuation relative to the mean. Thus, the value that we really care about is the *expected relative error:*

$$\text{relative error} = \frac{\text{standard error}}{\text{mean}} = \frac{\sqrt{pq}}{np}. \tag{8.23}$$

Consider two examples for a cell with $n = 100$ channels. If $p = 0.1$, most channels are closed and the relative error is 3%. If, on the other hand, $p = 0.9$ so that most channels are open, the relative error is tenfold smaller, only 0.3%.

The accuracy of photon detection by rods is also augmented by the stability of rhodopsin molecules. If rhodopsin had a tendency to change its form due just to thermal agitation, it could lead to false detection of photons. In reality, however, the *cis* isomer of rhodopsin is so stable that a single molecule has an even chance of maintaining its configuration for 420 years at mammalian body temperature! In other words, there is almost no chance that the false detection of a photon will be initiated by the spontaneous isomerization of any particular rhodopsin. There are, however, a very large number of rhodopsin molecules in each rod cell, with the net result that the cell mistakenly reports the arrival of a photon every couple of minutes in the dark (Schnapf and Baylor 1987). Fortunately, this rate is low enough that it does not interfere with human vision, even in very dim light.

But then, as animals go, humans are not very adept at seeing in the "dark." For animals adapted to very low light conditions (deep-sea fishes, for example), the accurate detection of just a few photons may be a matter of life and death. For example, the spontaneous isomerization of a rhodopsin molecule might appear the same to the visual system as the brief luminescent flash of a distant jellyfish. Flashing jellyfish can signal the approach of a predator. In a case such as this, spontaneous isomerizations could be the cause for considerable piscine anxiety and might set the practical limit to visual sensitivity. It probably helps that in addition to being dark, the deep sea is cold. The rate of spontaneous isomerization in rhodopsin decreases with a decrease in temperature. For a thorough discussion of the limits to sensitivity, consult Donner (1992).

8.6 *Noise and Ion Channels*

We now return to the subject of membrane potentials. The electrical potential across a membrane is used by living cells for a variety of purposes: the synthesis of ATP in mitochondria, the active transport of amino acids and sugars into the cell, and, as we have seen, communication with the central nervous system. In this last case, nerve cells in animals propagate information through the use of traveling waves of membrane voltage, waves that are known as *action potentials*. As with any kind of communication, the transmission of information by action potentials is subject to the exigencies of noise, and this noise is our next subject of consideration.

In our discussion of rod cells, we noted that sodium channels are subject to thermal agitation, and therefore can be expected to flicker open and closed. The probability of being in one state or another can be affected by a number of factors, one of which (the presence of cyclic GMP) we have discussed. Here we explore a second factor—these channels are sensitive to the membrane potential itself. At the resting potential of a nerve cell (as opposed to a rod cell), a sodium channel is biased toward being closed. This is a relatively stable situation. With the sodium channel closed, the voltage difference across the cell membrane is large, and the cell is *polarized*. With the membrane potential large, the sodium channel stays closed and the status quo is maintained.

If, however, the membrane potential decreases, the system can become unstable. As the voltage difference decreases, the probability is increased that sodium channels will be open, which allows for the influx of sodium. As we have seen, this current further depolarizes the cell membrane, which allows more channels to open, and the process feeds back on itself. The net result is that at a critical membrane potential (typically −40 mV) the system becomes unstable, and a drastic decrease in membrane potential is propagated along the cell membrane.

When an influx of sodium starts this feedback, an analogous chain of events is triggered in the cell's potassium channels, but the time course of the flux of potassium is slower than that of sodium. The combined result of both the influx of sodium and the subsequent efflux of potassium is the propagation of an action potential down the cell. The physics of action potentials is well known, and the fine points can be explored in any introductory text on neurophysiology. What concerns us here is not the full course of events, but rather just the initial influx of sodium that triggers the process.

We have already noted that sodium channels (like all molecules) are subject to thermal agitation. As a result, even when the cell's membrane potential is large and the channels are biased toward being closed, there is a certain chance that any given channel will open spontaneously. In most cases, such a chance opening would cause no problem; the resulting influx of sodium would not change the membrane potential sufficiently to start an action potential. As we will see, however, the smaller the cell, the larger the change in voltage caused by a single channel opening. Taking this thought to its logical conclusion, there must be some critical small size for nerve cells at which the chance opening of a single sodium channel is sufficient to trigger an action potential. For a cell of this size, thermal noise could pose a serious problem. How small does a cell have to be before the opening of a single channel can fire an action potential?

8.6.1 AN ELECTRICAL ANALOG

At this point, it is useful to translate the ideas shown in figure 8.4 into an equivalent electrical circuit as shown in figure 8.5. The potassium potential, V_K, is represented by a battery with its negative terminal toward the inside of the cell. That is, the effect of this battery is to drive the inside of the cell to a voltage

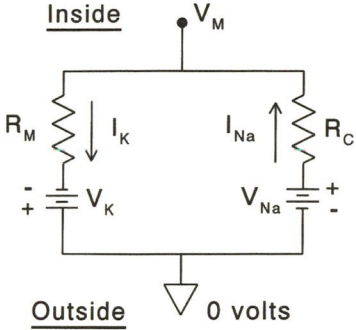

FIG. 8.5 A simple circuit that simulates the role of an ion channel in a cell membrane. See the text for an explanation.

that is negative relative to the outside. The "potassium battery" is connected to the rest of the circuit through a resistance, $\mathcal{R}_M$, that represents the resting permeability of the cell membrane to the flow of potassium ions.

The sodium potential, V_{Na}, is also represented by a battery, but in this case the positive terminal is toward the inside of the cell. This second battery is connected to the rest of the circuit by a resistance, $\mathcal{R}_C$, the resistance of the membrane to the flow of sodium ions. By tradition, the point in the circuit corresponding to the outside of the cell is set to 0V (ground), and the point in the circuit representing the inside of the cell is where we measure the membrane potential, V_M.[6]

From empirical experiments, neurophysiologists have shown that if V_M is more positive than -40 mV, an action potential is generated. Under what circumstances does this happen? We answer this question in two steps. First, we see how to calculate V_M from a knowledge of $\mathcal{R}_C$ and $\mathcal{R}_M$. We then calculate the size of the cell that gives us $\mathcal{R}_C$ and $\mathcal{R}_M$ at threshold.

8.6.2 CALCULATING THE MEMBRANE VOLTAGE

How can we calculate the membrane potential? To do so, we note two basic rules for electrical circuits. The current that flows (I, measured in amperes) is proportional to the difference in voltage, ΔV, between two points in the circuit, and inversely proportional to the resistance between these points, $\mathcal{R}$ (measured in ohms):

$$I = \frac{\Delta V}{\mathcal{R}}. \tag{8.24}$$

[6] For the circuit to be an accurate model of a real cell, we would have to include a capacitor in it, connected between ground and the point at which V_M is measured. This capacitor would give the circuit a time-dependent nature. For simplicity, however, we will not deal with time-dependent circuits here.

This is Ohm's law. Second, we note that in a circuit such as we have drawn here, electrical charges cannot be stored locally. In particular, the current, I_K, flowing away from the positive terminal of the potassium "battery" must be equal to the current, I_{Na}, flowing into the negative terminal of the sodium "battery" (see the arrows in fig. 8.5).

These two rules can be applied to our circuit with the following result:

$$\frac{V_M - V_K}{\mathcal{R}_M} = \frac{V_{Na} - V_M}{\mathcal{R}_C}. \tag{8.25}$$

This relationship can be solved for V_M in terms of the potassium and sodium potentials and their associated resistances:

$$V_M = \frac{V_K + (\mathcal{R}_M/\mathcal{R}_C)V_{Na}}{1 + \mathcal{R}_M/\mathcal{R}_C}. \tag{8.26}$$

So far we know what V_K and V_{Na} are, but we don't know either $\mathcal{R}_M$ or $\mathcal{R}_C$. This is where the fun begins. First consider $\mathcal{R}_M$, the resistance of the membrane to the flow of potassium ions. In a resting cell, there are always a few potassium ion channels are open, and the number of these channels sets the overall resistance. For a given type of cell at rest, the number of open channels is roughly proportional to the overall area of the cell membrane. In other words, if the spatial density of channels (channels per square meter) is constant, the smaller the membrane, the fewer channels there are. If at any time a fixed fraction of these channels open, the smaller the membrane, the fewer the number of open channels and the higher the resistance to the flow of potassium. This idea can be put in the form of an equation:

$$\mathcal{R}_M = \frac{\varrho_e}{A}, \tag{8.27}$$

where ϱ_e is the electrical resistivity of the resting membrane (ohms m^2) and A is the surface area of the cell. For a spherical cell of diameter d, $A = \pi d^2$, so

$$\mathcal{R}_M = \frac{\varrho_e}{\pi d^2}. \tag{8.28}$$

The resistivity of living cell membranes is highly variable, ranging from 10^{-4} to 10^2 ohm m^2 (Hille 1992), and we will take this variation into account. Inserting eq. (8.28) into eq. (8.26), we see that

$$V_M = \frac{V_K + [\varrho_e/(\pi d^2 \mathcal{R}_C)]V_{Na}}{1 + \varrho_e/(\pi d^2 \mathcal{R}_C)}. \tag{8.29}$$

What about $\mathcal{R}_C$, the resistance to the flow of sodium ions? In this case, our model is a bit simpler. For present purposes, we are interested in the effect of a single sodium channel on the membrane potential of the cell. When this channel is open, it has resistance $\mathcal{R}_C$, and a typical, single, open sodium channel has a resistance of about 4×10^{10} ohms. When our lone channel is closed, the only route for sodium to cross the membrane is shut off, and no sodium current can flow. Thus, when the channel is closed, $\mathcal{R}_C = \infty$.

8.6.3 CALCULATING THE SIZE

We are now in a position to use eq. (8.29) to calculate the diameter of a nerve cell as a function of its electrical properties. Solving for d, we find that the diameter is

$$d = \sqrt{\frac{\varrho_e}{\pi \mathcal{R}_C} \frac{V_{Na} - V_M}{V_M - V_K}}. \qquad (8.30)$$

In making our calculations, we use $V_K = -60$ mV, $V_{Na} = 60$ mV, $\mathcal{R}_C = 4 \times 10^{10}$ ohms, and a typical value of ϱ_e (11 ohms m^2). If we then set V_M to the critical membrane potential (-40 mV), we can calculate the *minimum* diameter the cell must have in order that a single channel opening will not fire an action potential. This predicted minimum size is 21 μm. Indeed, experimental measurements on rat olfactory receptor cells 20 μm in diameter ($\varrho_e = 11$ ohms m^2) have shown that these cells produce spontaneous action potentials in the absence of any perturbation other than thermal agitation (Lynch and Barry 1989).

So, small nerve cells are subject to the spurious firing of action potentials. Is there a way around this problem? If it were evolutionarily advantageous for animals to miniaturize their nerves, would physics allow it? The answer is both yes and no. Yes, an effective reduction in size could be accomplished by reducing the resting resistance of the cell membrane to the flow of potassium ions. For example, if $\varrho_e = 1$ ohm m^2 (instead of the 11 ohms m^2 used above), the minimum diameter of a nerve cell is reduced from 21 μm to 6.3 μm. If the membrane is made really leaky to potassium ($\rho_e = 10^{-4}$ ohm m^2), a cell could be only 0.063 μm in diameter before it encountered problems with spontaneous action potentials.

But the leakier the cell membrane, the harder the cell must work to maintain the membrane potential. As potassium leaks out of the cell, it is constantly being pumped back in by the action of sodium/potassium ATPases. As a consequence, the leakier the membrane, the more ATP must be hydrolyzed to maintain the status quo. This effect becomes increasingly severe as the cell gets smaller and its surface-to-volume ratio increases. In other words, yes, a small, effective nerve cell could be constructed, but only at a high metabolic cost.

8.7 *Noise and Hearing*

We now leave behind the realm of ion channels and shift to an exploration of the process by which we detect sound. Here, as with all the senses, there is a problem of noise. How can noise be a problem for hearing, you might think. After all, noise is what you hear! Actually, noise is what you do not want to hear. Recall that in this chapter we are dealing with noise in the sense of the disorder that accompanies the signal an organism is attempting to detect from its surroundings. Thus the "noise" of traffic or the annoying "noise" of a baby crying in a movie theater are, for our purposes, bona fide sounds. Instead, what concerns us here is the random, unpredictable noise caused by the thermal agitation of molecules.

Recall from chapter 7 that sound is a traveling wave of pressure. As a sound wave impinges on your eardrum, the pressure variation results in a variation of the force imposed on the tympanic membrane, the eardrum (see the anatomy outlined in fig. 8.6). As a result, the membrane deflects. It is this movement of the eardrum that is eventually transduced into the nervous impulses you "hear." Just as we limited our discussion of sight to the primary visual image created at the retina, we will limit the scope of our discussion of hearing. In this case, we will explore only the effects of pressure on the eardrum, and we will largely ignore the additional steps by which sound is mechanically transmitted to the inner ear, the process by which the sensory cilia of the cochlea

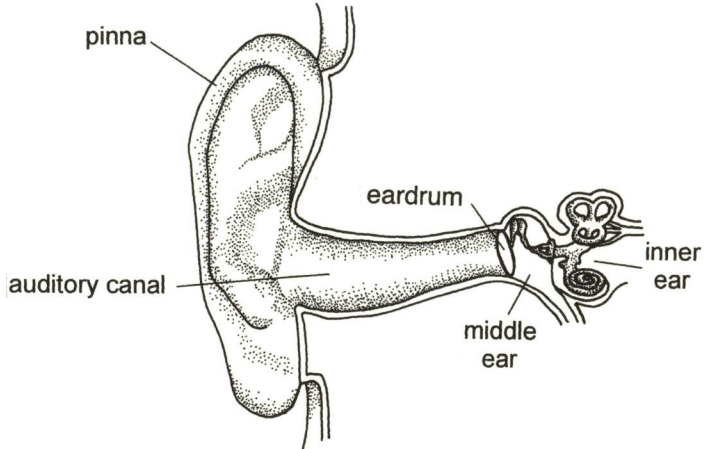

FIG. 8.6 The basic architecture of the human ear. The eardrum (tympanic membrane) deforms in response to fluctuations in pressure. This motion is transmitted by the bony linkage of the middle ear to the sensory apparatus in the inner ear.

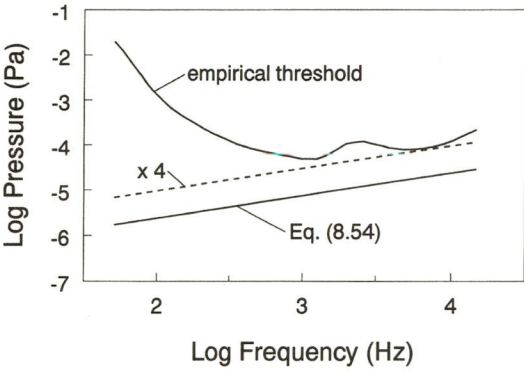

FIG. 8.7 The auditory sensitivity of humans varies with frequency (data from Killion 1978). We can sense the smallest fluctuation in pressure at a sound frequency of about 1000 Hz. Above this frequency, the auditory threshold roughly parallels the noise level predicted by eq. (8.54). If the eardrum were a quarter of its present diameter (a situation depicted by the dashed line), the noise level would approximately equal the auditory threshold.

transduce mechanical vibrations into nervous impulses, and all the processing of the nervous signals that goes on in the brain.

The first point we need to note about the sense of hearing is that it is incredibly sensitive. At a sound frequency of 1000 Hz, human ears can detect changes in pressure of approximately 2.9×10^{-5} newtons per square meter, a value equivalent to 0.29 *billionths* of an atmosphere (Rossing 1990). The area of a human eardrum is small, about 6.6×10^{-5} m^2, so the force required to result in "hearing" is only about 1.9×10^{-9} newtons, equivalent to the weight of 2×10^{-7} grams. In other words, if you turned your head on its side and someone carefully wiggled a dust mote weighing 0.2 micrograms on your eardrum, you could (in theory) notice it.

The sensitivity of human ears is a function of sound frequency (fig. 8.7). At low frequencies (below 1000 Hz), the ear is relatively insensitive; a large pressure fluctuation must be applied before the sound can be heard. Above 1000 Hz, the minimum noticeable pressure gradually rises with increasing frequency.

8.7.1 FLUCTUATIONS IN PRESSURE

Pressure as we sense it is due to the action of individual molecules as they bounce off a surface. As we have seen, at room temperature each molecule of gas in the air around us has a velocity of several hundred meters per second. This velocity, when coupled with a molecule's mass, means that each molecule has a certain momentum. When a gas molecule strikes a solid surface, it ricochets off. For example, consider a molecule that moves perpendicularly toward a stiff wall. When the molecule collides with the wall, its velocity is first brought to a halt, a process that (because it changes the molecule's momentum) requires a force to be exerted on the wall. The molecule then bounces back from where it came. Just as you apply a downward force to the floor when you jump up, the ricocheting molecule applies an additional force to the wall when it bounces back. It is the sum of all these molecular collisions that creates the pressure exerted by a gas. In any given interval, it is the average force exerted on an area of wall that we measure as the pressure.

As we know, molecules move at random due to thermal agitation. Therefore, we can guess that there will be some sampling error associated with the measurement of pressure. In other words, because pressure is a measure of the mean force exerted by molecules, there is very likely to be some fluctuation about the mean due to the random arrival of individual molecules. How big is this error? More to the point, given the extreme sensitivity of our ears, are these chance fluctuations in pressure large enough to affect our sense of hearing? Let's find out.

8.7.2 THE RATE OF IMPACT

There are two basic quantities we need to know before we can calculate the sampling error in pressure: the number of molecules that strike the eardrum in a given time interval, and the standard error in the momentum of these molecules when they hit. To estimate these values, we consider a chamber of arbitrary dimension as shown in figure 8.8. Here the right-hand wall of the chamber stands perpendicular to the x-axis and has area A. The chamber contains n gas molecules per volume, and these molecules move about at random. To begin, let's calculate how many molecules per second strike area A.

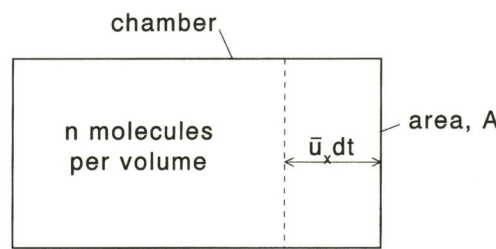

FIG. 8.8 A schematic diagram of the chamber used to calculate the number of molecules striking a wall. Note that the size and shape of the chamber are arbitrary; any chamber will do.

Because the motion of gas molecules is randomly directed, at any given time only about a third of the molecules in the chamber are moving predominantly in the x direction. In other words, at any specified time, approximately $n/3$ molecules per volume are moving more or less parallel to the x-axis. Of these $n/3$ molecules, half will be moving toward the right-hand end of the chamber, and half away. So we expect that roughly $n/6$ molecules per volume have their motion directed primarily toward A.

Now for some sleight of hand. If molecules move with a mean speed $\bar{u}$ (a value that is independent of direction), they will move a distance $\bar{u}dt$ in time dt. Thus, molecules within a distance $\bar{u}dt$ of area A and moving toward the wall will strike our area within time dt. But how many molecules is this? Well, the distance $\bar{u}dt$ multiplied by the area A is a volume, and we know the number of molecules per volume moving toward A ($= n/6$). Putting this all together, we see that the number of molecules that strike A is

$$\text{number striking } A \cong \frac{n}{6} \times A\bar{u}dt. \qquad (8.31)$$

From this we can calculate the approximate number of molecules per time that strike the walls of the chamber:

$$\text{number per time} \cong \frac{nA\bar{u}dt}{6dt} = \frac{nA\bar{u}}{6}. \qquad (8.32)$$

Note that this argument depends neither on the precise shape of the chamber nor its size, so we can assume that it will hold true for any area. Note also that we have been fast and loose with our assumptions in this calculation. In particular, we have glossed over the precise angles at which molecules move, so we cannot expect this calculation to be exact. However, when a more thorough analysis is carried out (see Reif 1965, 271–273), one finds that the exact answer is actually quite close to what we have calculated here:

$$\text{number per time, } N = \frac{nA\bar{u}}{4}. \qquad (8.33)$$

Thus, the number of molecules per time that strike an area (such as an eardrum) depends on the number of molecules per volume in the gas, the average speed of these molecules, and the area available to be struck.

8.7.3 FLUCTUATIONS IN VELOCITY

We can rearrange eq. (8.33) to our advantage. If we divide the number of molecules per time by n (the number of molecules per volume), we get the volume per time of gas that is effectively swept toward the wall. Dividing this volume per time by A, we end up with $\bar{u}_{x+}$, the average component of velocity parallel to the x-axis for those molecules moving toward the wall. In other words,

$$\bar{u}_{x+} = \frac{\bar{u}}{4}. \qquad (8.34)$$

We can immediately put this value to work in a calculation of the variation in the velocity with which molecules move toward A. By definition, for an arbitrary sample of molecules this variance is

$$\sigma^2(u_{x+}) = \frac{1}{j} \sum_{i=1}^{j} [u_{x+}(i) - \bar{u}_{x+}]^2, \tag{8.35}$$

where j is the number of molecules in the sample, and $u_{x+}(i)$ is the positive x-directed speed of the ith particle. Expanding the squared term gives us

$$\sigma^2(u_{x+}) = \frac{1}{j} \sum_{i=1}^{j} \left[u_{x+}^2(i) - 2u_{x+}(i)\bar{u}_{x+} + (\bar{u}_{x+})^2 \right]$$

$$= \left[\frac{1}{j} \sum_{i=1}^{j} u_{x+}^2(i) \right] - \left[2\bar{u}_{x+} \frac{1}{j} \sum_{i=1}^{j} u_{x+}(i) \right] + (\bar{u}_{x+})^2. \tag{8.36}$$

Fortunately, this expression can be simplified considerably. In the first term on the right of this equation, u_{x+} is squared. As a consequence, it really doesn't matter to this term that we have specified that molecules move in the positive x direction; if they moved in the negative x direction we would get the same answer. Thus,

$$\frac{1}{j} \sum_{i=1}^{j} u_{x+}^2(i) = \overline{u_x^2} = \frac{kT}{m}. \tag{8.37}$$

In the second term on the right of eq. (8.36), $\frac{1}{j}\sum_{i=1}^{j} u_{x+}(i)$ is just $\bar{u}_{x+}$. As a result, eq. (8.36) boils down to

$$\sigma^2(u_{x+}) = \frac{kT}{m} - 2(\bar{u}_{x+})^2 + (\bar{u}_{x+})^2$$

$$= \frac{kT}{m} - (\bar{u}_{x+})^2. \tag{8.38}$$

(Note that this is an example of the result we derived in chapter 4, where the variance is equal to the mean square minus the mean squared.) Making use of eq. (8.34), we conclude that

$$\sigma^2(u_{x+}) = \frac{kT}{m} - \frac{(\bar{u})^2}{16}. \tag{8.39}$$

By taking the square root of this variance, we arrive at what will turn out to be a useful expression: the standard deviation in the velocity of molecules as they move toward the wall:

$$\sigma(u_{x+}) = \sqrt{\frac{kT}{m} - \frac{(\bar{u})^2}{16}}. \tag{8.40}$$

8.7.4 FLUCTUATIONS IN MOMENTUM

Now for some basic physics. Momentum is the product of mass and velocity. Thus, in the positive direction along the x-axis the standard deviation of momentum of thermally agitated gas molecules is

$$\sigma \text{ of momentum} = \text{mass} \times \sigma \text{ of velocity}$$

$$= m\sqrt{\frac{kT}{m} - \frac{(\overline{u})^2}{16}}. \tag{8.41}$$

As we have noted, when a molecule bounces perpendicularly off a surface, it is first brought to a halt (requiring a change in momentum equal to the original, mu_{x+}) and then propelled backward (again requiring a change equal to the original). The net change in a molecule's momentum in colliding with a surface is thus $2mu_{x+}$. This momentum is supplied by the wall. Applying this principle to the problem at hand, we see that the standard deviation of momentum imparted to the wall by each impacting molecule is

$$\sigma \text{ of momentum imparted to the wall} = 2m\sqrt{\frac{kT}{m} - \frac{(\overline{u})^2}{16}}. \tag{8.42}$$

8.7.5 THE STANDARD ERROR OF PRESSURE

We now know what we originally set out to calculate: the number of molecules striking a wall per time (eq. 8.33) and the standard deviation of the momentum imparted by individual molecules (eq. 8.42). Where has it gotten us? By noting that momentum per time is equal to force (Newton's second law), we find ourselves in a position to calculate the standard deviation of the force applied to a given area of wall by the molecules in a gas:

$$\sigma \text{ of force} = \frac{\text{number of molecules impacting}}{\text{time}} \times \frac{\sigma \text{ of momentum}}{\text{molecule}}$$

$$= 2Nm\sqrt{\frac{kT}{m} - \frac{(\overline{u})^2}{16}}. \tag{8.43}$$

But this is the standard deviation of the force exerted by individual molecules. What we are interested in here is the standard deviation of the *mean* force for a sample of several molecules, because it is this average force that determines the pressure. But the standard deviation of the mean is, by definition, the standard error:

$$\text{standard error of force} = \frac{\sigma \text{ of force}}{\sqrt{\text{number of molecules sampled}}}, \tag{8.44}$$

a value we could calculate if we knew the number of molecules in a sample. Fortunately, this is a value we can easily obtain. We know that the number of molecules hitting area A per time is N, so the number of molecules whose momentum is "sampled" by the area is Nt. But it is not just one sample that concerns us; it is the variation among repeated samples that is the issue here. As a consequence, it is better for us to imagine taking a series of samples, each of interval t. In that case, we are sampling at a frequency $f = 1/t$, and the number of molecules hitting area A in each of our samples is N/f. Thus, the standard error of force is

$$\text{standard error of force} = \frac{2Nm\sqrt{\frac{kT}{m} - \frac{(\overline{u})^2}{16}}}{\sqrt{\frac{N}{f}}} = 2m\sqrt{Nf\left(\frac{kT}{m} - \frac{(\overline{u})^2}{16}\right)}. \quad (8.45)$$

Dividing force by area, we can at long last calculate the standard error of the pressure exerted on the eardrum:

$$\text{standard error of pressure} = \frac{2m\sqrt{Nf\left(\frac{kT}{m} - \frac{(\overline{u})^2}{16}\right)}}{A}. \quad (8.46)$$

This is very close to the answer we seek. To arrive at a quantitative answer, however, we need values for $N, \overline{u}$, and A, and to calculate N we also need to know n.

8.7.6 QUANTIFYING THE ANSWER

First, we define the area over which we measure the pressure. An eardrum is roughly circular, and it has two faces on which molecules can impact. Because molecules act independently on the different faces, we can add the two areas to calculate the total area acted upon by sound pressure. Thus, if the eardrum has radius r,

$$A = 2\pi r^2. \quad (8.47)$$

We now need to calculate n, which, given eq. (8.33), will allow us to calculate N. To accomplish this, we ask you to harken back to your highschool chemistry class and to recall a result from the ideal gas law:[7]

$$\mathcal{PV} = \mathcal{N}kT. \quad (8.48)$$

---

[7] Here we express the combined gas law in terms of individual molecules; it is more often seen expressed in terms of moles of molecules. In that form, the number of molecules is replaced by the number of moles, and k (Boltzmann's constant) is replaced by R (the gas constant, the product of Avogadro's number and k).

Time averaged pressure ($\mathcal{P}$) times volume ($\mathcal{V}$) is equal to the product of $\mathcal{N}$ (the total number of molecules present), Boltzmann's constant, and absolute temperature. Noting that $\mathcal{N}/\mathcal{V} = n$, we see that

$$\mathcal{P} = nkT. \tag{8.49}$$

Thus,

$$n = \frac{\mathcal{P}}{kT}. \tag{8.50}$$

In other words, we can now calculate n in terms of atmospheric pressure and temperature, both of which are easily measured.

Finally, we need to know the average molecular velocity, $\bar{u}$. The full calculation of $\bar{u}$ follows from a consideration of the random directions in which molecules move, but the process is best left to a text on statistical mechanics (see Reif 1965, 268, for example), and we borrow the result without derivation:

$$\bar{u} = \sqrt{\frac{8kT}{\pi m}}. \tag{8.51}$$

We can now insert these expressions for A, n, and $\bar{u}$ into eq. (8.33) to calculate N, and then substitute N and $\bar{u}$ into eq. (8.46). When the dust settles, we see that the standard deviation of the pressure imposed on the eardrum (that is, the standard error of the mean force per area exerted by impacting molecules) is

$$\text{standard error in pressure} = \sqrt{\mathcal{P}\sqrt{\frac{32}{\pi}}\left(1 - \frac{1}{2\pi}\right)\sqrt{mkT}} \times \sqrt{\frac{f}{\pi r^2}}. \tag{8.52}$$

So, for a given temperature and gas (the type of gas sets the weight per molecule, m) at atmospheric pressure (which sets $\mathcal{P}$ to 1×10^5 N m^{-2}), the variation in pressure on the eardrum is proportional to $\sqrt{f/(\pi r^2)}$. The smaller the radius of the eardrum, the smaller the number of molecules sampled in a given time, and the larger the sampling error. Similarly, the shorter the time interval in a sample (that is, the higher the frequency), the smaller the number of molecular collisions sampled, and the larger the potential variation in pressure.

For nitrogen molecules (the principal component of air, $m = 4.7 \times 10^{-26}$ kg) at one atmosphere ($\mathcal{P} = 10^5$ N m^{-2}) and an absolute temperature of 290 K (room temperature),

$$\sqrt{\mathcal{P}\sqrt{\frac{32}{\pi}}\left(1 - \frac{1}{2\pi}\right)\sqrt{mkT}} = 1.95 \times 10^{-9} \text{ kg s}^{-3/2}. \tag{8.53}$$

As noted previously, the area of a human eardrum is about 6.6×10^{-5} m^{-2}. Thus, the standard error of pressure acting on a human eardrum is

$$\text{standard error in pressure} = 2.4 \times 10^{-7}\sqrt{f} \quad \text{N m}^{-2}. \tag{8.54}$$

This relationship is graphed in figure 8.7. For sound frequencies greater than 1000 Hz, the threshold of human hearing is above and approximately parallel to the relationship of eq. (8.54). It is tempting to think that for this range of frequencies the auditory threshold has been adjusted through the course of evolution to stay safely above the level of thermal noise. If this is true, however, the safety factor is quite large. The gap between eq. (8.54) and the threshold of hearing is about three standard errors, indicating that only about one pressure wave in a thousand randomly exceeds the threshold.

Note again, however, that the standard error of pressure varies inversely with the radius of the eardrum. As a result, the smaller the eardrum, the worse the noise at any given frequency. For example, if human eardrums were one quarter the diameter they are, the pressure variations would be four times those calculated here (the dashed line in fig. 8.7) and approximately equal to the auditory threshold. This effect might help to explain why eardrum size appears to scale differently with body size than does the size of other organs. For example, small

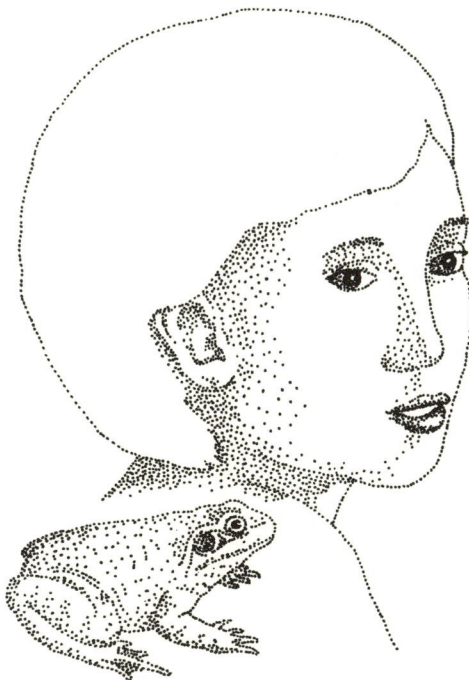

FIG. 8.9 A comparison of the size of a human and a bullfrog. Even though the frog is much smaller, its tympanic membrane (the dark, stippled area behind the eye) is roughly the same size as a human eardrum.

lizards and frogs have tympanic membranes almost the size of human eardrums (fig. 8.9). If the "eardrums" in these animals were small in direct proportion to their diminutive body length (that is, if animals scaled isometrically), the resulting small area would be prone to increased thermal noise.

8.8 *The Rest of the Story*

The calculations we have just made are only part of the story, and the plot thickens with further scrutiny. For example, as we noted previously, the sonic vibrations of the eardrum are transmitted to the fluid of the cochlea, where they result in the displacement of sensory cilia. We have already seen that these cilia should be subject to thermal agitation, and a more thorough analysis of the mechanics of the sensory cilia suggests that the root mean square displacement of the cilia should be approximately 10^{-9} m (Bialek 1987; Block 1992). It is surprising, then, when calculations show that the human ear can reliably detect sounds that result in displacements of the cochlear cilia of only 10^{-10} to 10^{-11} m, one to two orders of magnitude less than the displacements caused by thermal agitation.

The process by which our ears manage to detect faint sonic signals in the midst of thermal noise is an area of active research. Present understanding suggests that several mechanisms are involved. First, the sensory cilia are capable of active motion, and it appears that this capability can be used to filter out noise that does not occur at a particular frequency specific to each cilium. Second, there is much postreception processing of the nervous signals transmitted from the ear. The brains of animals are extraordinarily good at making sense of complex signals, and scientists have only begun to understand how they do it.

8.9 *Stochastic Resonance*

Up to this point, we have treated noise as something that is unavoidable, and a necessary evil with which sensory systems must cope. There are cases, however, in which random noise can actually be an important asset to a sensory system, allowing it to detect signals that would otherwise be undetectable. In fact, for certain systems, the more noise you add (up to a certain limit), the stronger the signal comes through. These systems exhibit what is known as *stochastic resonance*.

8.9.1 THE UTILITY OF NOISE

The idea behind stochastic resonance is actually quite intuitive. Consider the simple mechanical system shown in figure 8.10, a crude design for a tiltmeter.

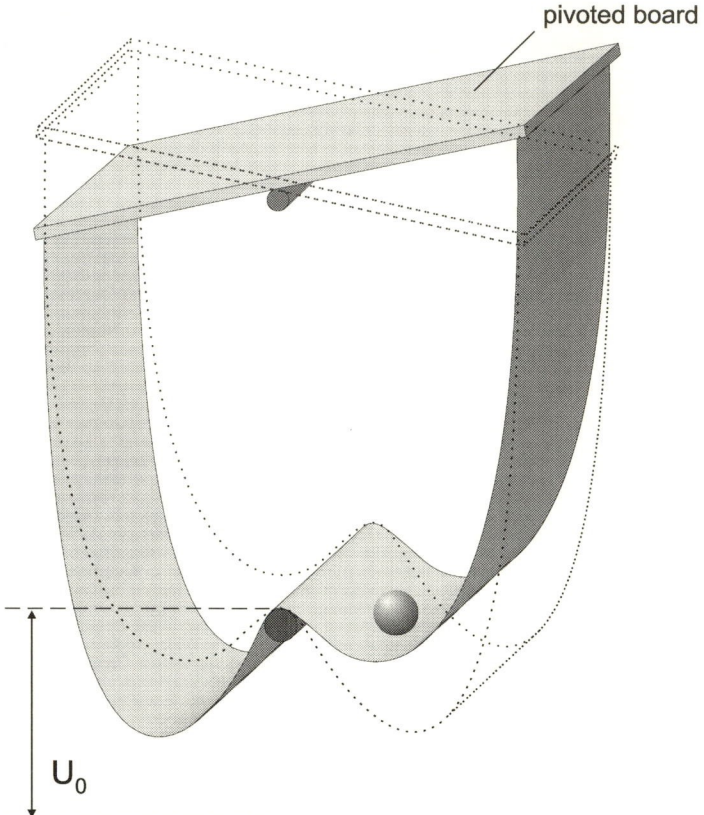

pivoted board

U_0

FIG. 8.10 A schematic diagram of the simple two-well tiltmeter described in the text. Given sufficient energy, the ball will switch wells in time with the tilt of the board.

A length of flexible material is draped from a pivoted board. The material is divided into two wells separated by a low barrier, and a ball is placed in the well on the left. If the board is tilted to the right, the material in the left-hand well is lifted, and the ball begins to roll. If the tilt is sufficient and the ball has enough momentum, the ball can cross over the barrier into the well on the right. As the board sways back and forth, it is possible for the ball to switch wells in time with the board's motion.

The location of the ball could thus be used as a reliable indicator of the board's tilt. When the ball is in the right-hand well, the board is indeed tilted to the right; presence of the ball in the left-hand well indicates that the board is tilted to the left. The frequency with which the ball switches wells is a good measure of the frequency at which the board sways. In other words, when the signal applied to the system (the tilt) is energetic enough, it can be reliably detected by the location of the ball. None of this should be surprising.

The surprise comes when we consider a situation in which the tilt of the board is too slow or too small to cause the ball to switch wells. In this case, the tilt signal is seemingly too weak to be detected by our crude meter. There is a fix, however. Let us apply some mechanical noise by shaking the whole apparatus up and down randomly. Normally one would think that noise would serve only to confound the detection of a weak signal, but in this case it does just the opposite. The random shaking of the board stirs up the ball, occasionally giving it enough momentum to cross the barrier into the other well. This transfer is most likely to occur when the system is already predisposed for the ball to switch wells. In other words, the random increase in momentum provided by noise is most likely to cause the ball to move from the left-hand well to the right-hand well when the board is tilted to the right. There are no guarantees, mind you. An energetic boost may be applied to the ball at an inauspicious time, causing the ball to switch wells contrary to the tilt of the board, or the ball will not switch wells when it is predisposed to do so because, by chance, the necessary boost may not be applied. But *on average* the ball will switch wells in time with the board's tilt. Thus, in the presence of random noise, the location of the ball can be used to estimate the frequency with which the board is tilted, even when the tilt would otherwise be undetectable. This is what is meant by the term "stochastic resonance." The system is resonant in the sense that the ball switches wells more or less in time with the tilt signal, but the process is stochastic because it relies on noise.

A moment's thought should bring you to the conclusion that there are limits to the efficacy of stochastic resonance. If the level of noise is too small, the ball seldom if ever will get the boost it needs to cross the barrier between wells, and a weak tilt signal will not be reliably detected. On the other hand, if the noise is too energetic, the ball will switch wells largely at the whim of the noise alone, and the resonance with tilt will be lost. There should be some discrete range in which the noise level serves to amplify the detectability of the tilt signal. This is indeed true.

Before we quantify this notion, we need to digress for a moment to consider how to measure the reliability with which a system detects a signal, a characteristic traditionally expressed as the "signal-to-noise ratio." The idea here is that any detection system will fluctuate through time, and therefore will have a measurable variance in its state. A variety of indices can be used to describe the state of a system. In this case, it is convenient to use the gravitational potential energy of the ball. For example, if the ball has mass, its gravitational potential energy (mass × gravity × height) varies depending on its location. The bottom of each well goes up and down as the board tilts, so the ball's potential energy varies somewhat even if it does not change wells. On the other hand, the ball's gravitational potential energy is very high as it crosses the barrier between wells and very low as it settles into the lower of the two wells. Thus, switching wells entails a large shift in potential energy.

Now, we can quantify the fluctuation in the ball's gravitational potential energy by sampling the location of the ball at random times and calculating the standard deviation. Some of the ball's fluctuations are in time with the underlying signal provided by the board's tilt, and the standard deviation of these in-phase fluctuations is a quantitative measure of the magnitude of our *signal*. In contrast, some fraction of the overall variation is due to pure stochastic processes as the ball is randomly boosted from one well to the other. The standard deviation of these random fluctuations is a measure of the system's *noise*. The signal-to-noise ratio is defined as the ratio of these two standard deviations.

Now back to stochastic resonance. It has been shown for a wide variety of systems (see the reviews by Wiesenfeld and Moss 1995; Bulsara and Gammaitoni 1996) that the signal-to-noise ratio varies according to the following relationship:

$$\text{signal-to-noise ratio} \propto \left(\frac{\lambda_s}{\lambda_n}\right)^2 \exp\left(-\frac{U_0}{\lambda_n}\right). \qquad (8.55)$$

Here λ_s is the strength (=standard deviation) of the signal (measured in the absence of noise), λ_n is the intensity of the noise (measured in the absence of a tilt signal), and U_0 is the energy required to cross the barrier between wells. This relationship is shown in figure 8.11. Below some critical noise intensity, the signal-to-noise ratio is very small. As the intensity of noise increases above the critical point, however, the ratio rises rapidly to a peak. At this maximum, the noise is just sufficient to allow the ball to switch wells in time with the underlying signal and the signal is most reliably detected. If the intensity of noise is increased further, the signal-to-noise ratio falls because the alternation between wells is driven more and more by the noise than by the signal. The existence in a system of this type of relationship between the intensity of noise and the signal-to-noise ratio is evidence for the presence of stochastic resonance.

8.9.2 Nonlinear Systems

What kind of systems can exhibit stochastic resonance? The basic requirements are twofold. First, the frequency of the signal that is to be detected must be small compared to the frequency of the noise. For example, in our tiltmeter, resonance is achieved reliably only if the noise has several chances to boost the ball over the barrier each time the board is tilted. In contrast, if noise boosted the ball only once in several oscillations of tilt, the frequency of the tilt signal

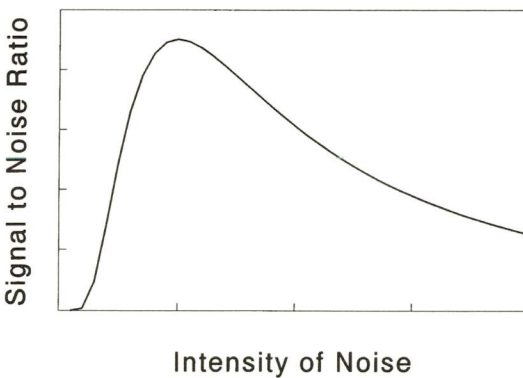

FIG. 8.11 The theoretical relationship between the noise added to a weak signal and the signal-to-noise ratio of the detected signal (eq. 8.55).

could not be reliably detected. The second requirement is that the system be *nonlinear* in its response. To see what this means, consider our tiltmeter with the board horizontal and the ball in the left-hand well. If the board is tilted to the right just slightly, the ball stays in its current well, and the change in its gravitational potential energy is small. A slightly larger tilt produces only a slightly greater change in energy. At some critical tilt, however, the ball jumps to the right-hand well, and, in the process moves to a much lower potential energy. In other words, the response of the ball is not linearly related to the tilt of the board—the system is nonlinear.

This type of nonlinearity is especially typical of systems that have a threshold above which some action occurs. Nerve cells are an excellent biological example, and they are capable of stochastic resonance. Consider the following scenario. As long as the membrane potential of a hypothetical nerve cell remains more negative than some threshold value (−40 mV, say), the cell is quiescent. If the membrane voltage rises above this threshold, however, an action potential is propagated down the cell body. What happens if we now apply a faint voltage signal to the cell? For the sake of argument, let's assume that the variation in this signal is such that it never changes the membrane voltage sufficiently to fire an action potential. Under these circumstances, the faint signal is undetectable by the nerve cell.

We now add some random noise to our faint voltage signal. If the amplitude of the noise is large enough, sometimes when the periodic signal is nearest the voltage threshold, the additional voltage due to the random noise will cause the membrane potential to exceed the threshold value, and the cell fires. As with the ball-and-wells system, the response of the nerve is not perfect. Sometimes the random noise causes the nerve cell to fire even when the periodic signal is not near its peak, and sometimes when the periodic signal is near its peak, the instantaneous noise level is insufficient to exceed the threshold. But, on average, the cell will fire an action potential at a time near when the faint signal is at its peak, and the pattern of action potentials in the cell will be closely coupled to the frequency of the periodic signal. In other words, given sufficient noise, the nerve cell can detect an otherwise undetectable periodic signal.

8.9.3 THE HISTORY OF STOCHASTIC RESONANCE

The theory of stochastic resonance has an interesting history. It began with a question as to why some aspects of Earth's climate (in particular, the ice ages) seem to occur in step with some extremely small deviations in Earth's orbit. The periodicity in the ice ages had been known by geologists and paleontologists since 1930 (Milankovitch 1930), although a plausible explanation for the phenomenon had not been worked out. In 1982, a group of physicists had the innovative thought that because Earth's climate seems to exhibit the type of nonlinear, bi-stable behavior we have modeled with our tiltmeter (ice age and no ice age being the two "wells"), random fluctuations in climate might be sufficient to amplify the weak variations in solar insolation due to Earth's changing orbit (Benzi et al. 1982; Nicolis 1982). Thus the study of stochastic resonance was born, and it has been an active area of research ever since.[8]

Stochastic resonance has been detected in a variety of electronic and mechanical systems, as well as in the sensory systems of crickets (Levin and Miller 1996), crayfish (Douglass et al. 1993), and humans (Inglis et al. 1996). Stochastic resonance can even occur at the level of individual ion channels (Bezrukov and Vodyanoy 1995). More recently, computer models of multineuron systems have suggested that the conditions under which stochastic resonance occurs may be less restrictive than previously thought. The existence of a strict threshold in the response of a system may not be required (Bezrukov and Vodyanoy 1997), and the necessary "tuning" of the noise level to the response of the system may not be as strict as implied by eq. (8.55). For example, Collins et al. (1995) have shown that by combining many model neurons into a network, a weak signal can be reliably detected across a wide range of noise levels, and Moss and Pei (1995) describe a similar phenomenon for a collection of ion channels.

These last two findings are of particular interest to neurophysiologists, who for many years have known that the nervous systems of animals are inherently noisy. Ever since the discovery of stochastic resonance, it had been tantalizing to think that this nervous noise could actually serve to amplify the system's ability to detect weak signals. But the basic theory of stochastic resonance seemed to imply that to achieve reliable detection, the level of noise would have to

---

[8] There are few new ideas under the sun (or in orbit around it). Once the utility of stochastic resonance was made apparent to scientists, its roots in earlier work were pointed out. Dykman and McClintock (1998) note that stochastic resonance is an extension of calculations performed by Debye in the 1920s related to the reorientation of polar molecules.

be adjusted continuously as the strength of a signal varied. Evidence for this "tuning" had not been found, leaving in limbo the applicability of stochastic resonance to nervous systems. The findings of Collins et al. (1995) and Moss and Pei (1995) largely remove the need for tuning, with the result that nervous noise may again be thought of as a boon to (rather than the bane of) signal detection in animals.

Russell et al. (1999) provide the first tentative evidence that stochastic resonance actually works to increase the fitness of an animal in the wilds. The paddle fish (*Polyodon spathula*) uses passive electroreceptors in its rostrum (the "paddle") to detect the minute electrical signals produced by its prey, the water flea, *Daphnia*. The introduction of the proper amount of electrical noise into the water near the paddle fish substantially increases the distance over which the fish catches individual prey. In some cases, the rate at which prey are caught is doubled in the presence of the optimal amount of noise! Russell *et al.* suggest that in nature swarms of *Daphnia* themselves produce sufficient noise to enhance the feeding ability of the paddle fish.

The utility of stochastic resonance should not be overblown. Stochastic resonance can (under the appropriate circumstances) enhance the *detectability* of a weak signal, but because noise is added in the process, *the detected signal will be noisier than the underlying signal itself*. Claims that stochastic resonance can actually enhance the signal-to-noise ratio of the detected signal relative to the underlying signal must therefore be viewed with suspicion (Dykman and McClintock 1997).

8.10 *Summary*

Through the course of evolution, the sensory systems of animals have evolved exceptional sensitivities—the ability to detect single photons, for instance, or the passage of a few ions across a membrane. Given these sensitivities, organisms are commonly asked to cope with the consequences of unavoidable noise due to thermal and quantum effects. These consequences can serve to set limits on sensation: our eyes can detect the fine details in an image, but only if sufficient time is taken; our ears can hear the barest whisper, but only if our eardrums are large enough. These trade-offs among size, time, and signal detection give new insight into the design of sensory systems. Indeed, recent research has revealed ways in which our senses seemingly outsmart physics (using random noise to help enhance the detectability of weak signals, for example), and research into the mechanics of perception remains an active and intriguing field. The interested reader may wish to consult Berg and Purcell (1977) and Bialek (1987).

8.11 *A Word at the End*

In these eight chapters we have covered a lot of intellectual territory. In chapter 2 we began with the basic concepts and rules of probability theory, and in the course of exploring the algebra of chance discovered a variety of situations in which our intuition could lead us astray. Who would guess that with a mere twenty-three people in a room there is an even chance that two have the same birthday? Or that even a minuscule chance of a false positive can severely jeopardize our ability to test for a rare trait? Why should it be more likely that a gambler will get at least one "1" in six throws of a die than two "1's" in twelve throws? The message we hope is clear: in a stochastic world, spurious (and perhaps costly) conclusions can be avoided only through the careful and logical application of the theory of probability. In chapters 3 and 4 we built on our understanding of the laws of probability to develop the concept of a probability distribution. By associating each possible value of a random variable with its probability of occurrence, we formulated a compact and efficient method for describing all that can be known about a stochastic process. We derived the binomial and geometric probability distributions and the uniform and normal probability density functions. At that point, the definitions, theorems, and rules were probably getting a bit cluttered in your mind, but we hope that in retrospect their utility has become apparent. The Central Limit Theorem really does work! Its assurance that the sum of random variables is normally distributed has allowed us to explain the vast panoply of processes explored in chapters 5 and 6: random walks in molecular and turbulent diffusion, the efficiency of ion channels and receptors, genetic drift, and rubber elasticity. In chapter 7, we even managed to use the statistics of Gaussian processes to predict the probability that our ear drums will be damaged at a cocktail party, and through an extension of that logic we have predicted the maximum age of human beings and the return of the 0.400 hitter in baseball. In many ways, chapters 5–7 were devoted to a description of the many types of "noise" that plants and animals encounter in the world. In this final chapter we have examined how biology has coped with this inevitable noise: the trade-off between visual acuity and response time, the issue of scale in the evolved design of ears. We have even seen how noise can enhance the detection of weak signals in our environment.

At this point we have come full circle. Our basic premise has been that the inevitability of chance is a fact that should be embraced by biologists in their study of life. With the suggestion that nerves may actually *need* some noise to function effectively, we have come about as far as we can to convince you of the utility of chance in biology. So, at this point, we turn the exploration over to you. We hope that we have piqued your curiosity. If we have done our job, you should have a new glint in your eye as you view the randomness around you, a desire to understand how it all works, and the tools to make that understanding possible.

8.12 *A Problem*

You work for the Acme Microelectronics Group, and you have been asked to design an ultraminiature, ultrasensitive inclinometer for a geophysicist who studies earthquakes. The function of the machine is to measure small changes in the inclination of a bit of rock as pressure builds up prior to a quake. You decide to use a simple pendulum as the basis for your design. In essence, you plan to construct a tiny platform that can be glued to the rock with its surface initially horizontal. You will then hang a pendulum from the platform (see fig. 8.12). The acceleration due to gravity acts to insure that the pendulum shaft stays vertical even if the platform tilts, and a sensitive, noise-free electronic device is coupled to the pendulum's base to measure the angle between the pendulum's shaft and the normal to the platform. This angle, θ, is zero when the platform is horizontal, and increases with increasing tilt. You desire to make measurements accurate to a tenth of a microradian (10^{-7} radians). Simple!

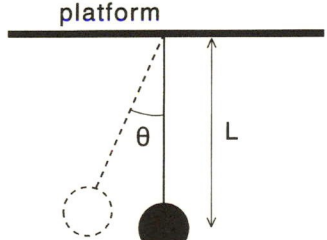

Fig. 8.12 A schematic diagram of the Acme inclinometer.

All is well until you try to make a device with a pendulum length only a few microns long. It seems that the smaller you make the meter, the noisier it gets. From your newfound knowledge of thermal noise, calculate and graph how the inherent noise of the device varies with the length of the pendulum and the weight on its end.

The following equations will come in handy. For small θ,

$$\text{restoring force} = mg\theta,$$

$$\text{distance moved by the mass} = L\theta,$$

where m is the mass at the end of the pendulum, g is the acceleration due to gravity (9.81 m s^{-2}), and L is the length of the pendulum. The length of the pendulum can be changed separately from the mass. Recall that energy is the integral of force times the distance through which the force moves its point of application.

After fiddling with the device for a while, you find that you can increase the accuracy of your angular measurement by averaging values over time. The natural frequency of a pendulum's oscillation is (to a first approximation):

$$\text{natural period} = 2\pi \sqrt{\frac{L}{g}}.$$

Let's assume that all the thermal fluctuations in the pendulum's angle occur at the pendulum's natural frequency, and that you can thus take one independent measurement of angle per the period with which the pendulum oscillates. Calculate how the accuracy of your measurement of θ varies with measurement time for pendula of various sizes (both mass and length). Discuss the trade-offs between the temporal response of your meter and the mass and length of its pendulum.

8.13 *Appendix*

Here is an informal derivation of the Poisson distribution (after Ruhla 1992). We begin (as always) with the binomial distribution:

$$P(i|n, p) = \frac{n!}{i!(n-i)!}p^i q^{n-i},$$

where i is the number of successes in n trials, with p being the probability of success in any one trial. We would like to find an approximation to this distribution that is valid when the probability of success is small (that is, when $p \ll 1$). If the probability of success is small, the number of successes will be much smaller than the number of trials, so we also may assume that $i \ll n$. For reasons that will become apparent, we also need to assume that n is large.

With these conditions in hand, let's expand the first term in the binomial distribution:

$$\begin{aligned}
\frac{n!}{i!(n-i)!} &= \frac{1 \times 2 \times \ldots \times (n-i-1) \times (n-i) \times (n-i+1) \times \ldots \times (n-1) \times n}{i!(n-i)!} \\
&= \frac{(n-i)! \times (n-i+1) \times \ldots \times (n-1) \times n}{i!(n-i)!} \\
&= \frac{(n-i+1) \times \ldots \times (n-1) \times n}{i!}.
\end{aligned} \tag{8.56}$$

We now make the first use of our conditions. If $i \ll n$ and n is large, then $(n - i + 1) \cong n$. The same is true for each of the values multiplied in the numerator of eq. (8.56). There are a total of i of these values, leading us to the conclusion that

$$\frac{n!}{i!(n-i)!} = \frac{(n-i+1) \times \ldots \times (n-1) \times n}{i!} \simeq \frac{n^i}{i!}. \tag{8.57}$$

Now for the second expression in the binomial distribution. Because $p+q = 1$, we know that $q = 1 - p$. Thus,

$$p^i q^{n-i} = p^i (1-p)^{n-i}. \tag{8.58}$$

Again invoking our conditions, we note that $i \ll n$, so that $(1-p)^{n-i} \simeq (1-p)^n$. Now for some mathematical sleight of hand:

$$(1-p)^n = \exp[n \ln(1-p)]. \tag{8.59}$$

But if $p \ll 1$, $\ln(1-p) \cong -p$ (try it!). As a result,

$$p^i(1-p)^n \cong p^i \exp(-np). \tag{8.60}$$

Combining eq. (8.57) with eq. (8.60), we find that

$$P(i) = \frac{n^i p^i}{i!} \exp(-np)$$
$$= \frac{(np)^i}{i!} \exp(-np). \tag{8.61}$$

We are almost there. It remains only to recall from chapter 4 that the product np is the mean of the binomial distribution. Substituting μ for np in eq. (8.61), we arrive at our destination:

$$P(i) = \frac{\mu^i}{i!} \exp(-\mu), \tag{8.62}$$

subject to the conditions that the chance of success is small and the number of trials is large. This is the Poisson distribution, as advertised.

9

The Answers

9.1 *Chapter 2*

1. Enumerate the sample space. There is only one gesture in which all five fingers are folded; similarly, there is only one gesture in which all fingers are raised. There are five gestures in which a single finger is raised and (through a consideration of the complement of each gesture) five in which a single finger is folded. With a bit of fiddling you should be able to convince yourself that there are ten different gestures in which two fingers are raised and three folded and another ten in which three are raised and two folded. In all, there are thirty two distinct hand gestures. The probability that your random choice of a gesture matches the sole insulting gesture in your rescuer's culture is thus $1/32 = 0.031$. An efficient way to calculate the number of different gestures (a formula for the number of combinations) is discussed in chapter 3.

2. There are two possible outcomes for each birth (boy, b, or girl, g). As a result, there are $2^4 = 16$ possible outcomes in the four births. Of these, only six have equal numbers of each gender ($bbgg$, $bgbg$, $bggb$, $ggbb$, $gbgb$, $gbbg$). Thus, the probability of having an equal number of boys and girls in four children is $6/16 = 0.375$. In contrast, there are four ways of having one boy and three girls and four ways of having three boys and one girl, so the probability of ending up with three kids of one sex and one of the other is $8/16 = 0.5$. The probability of having either all boys or all girls is $2/16 = 0.125$ (a fact at which you could arrive either by logic or by subtraction).

3. First calculate the probability that gambler 1 does *not* throw at least one ace. The probability of not throwing an ace on any given throw is $5/6$. The probability of repeating this six times is $(5/6)^6$. This, then, is the probability of throwing zero aces. The probability of throwing at least one ace is thus $1 - (5/6)^6 = 0.6651$ (rule 2).

We can proceed the same way for gambler 2. The probability of throwing zero aces in twelve tries is $(5/6)^{12}$. The probability of throwing an ace on the first throw, followed by eleven non-aces, is $(1/6) \cdot (5/6)^{11}$. There is the same probability if the single ace is thrown on the second throw, or third, or fourth, etc. Therefore, the probability of throwing exactly one ace in twelve tries is $(1/6) \cdot (5/6)^{11} \cdot 12$ (an application of rules 1 and 6). The probability of throwing either zero or one ace is $(5/6)^{12} + (1/6) \cdot (5/6)^{11} \cdot 12 = 0.3813$. The probability of throwing two or more aces in twelve tries is thus $1 - 0.3813 = 0.6187$. It is slightly more probable that gambler 1 throws at least one ace in six throws than it is that gambler 2 throws at least two aces in twelve throws.

4. This problem involves the product of a series of independent conditional probabilities. Consider the population with eight deer, four of which are marked. The probability is 4/8 that the first deer captured is marked. Given that this occurs, the only way that we can capture four more deer (for the prescribed total of five) and still only have one marked deer is if all the remaining deer are unmarked. After the first (marked) deer has been caught, there are four unmarked deer in the remaining population of seven. Thus, the probability that the second deer captured is unmarked is 4/7. There are then three unmarked deer in the remaining population of six, so the probability that the third deer captured is unmarked is 3/6. Continuing this process for the fourth and fifth deer, we see that when that marked deer is caught on the first try, the overall probability of catching exactly one marked deer is (from the product rule):

$$\frac{4}{8} \cdot \frac{4}{7} \cdot \frac{3}{6} \cdot \frac{2}{5} \cdot \frac{1}{4} \cong 0.0143. \tag{9.1}$$

But we could also catch our sole, marked deer on the second try. Working through the math we find that the chance of this occurring is also 0.0143. The same for catching the marked deer on the third, fourth, or fifth tries. The overall probability of catching exactly one marked deer is therefore $5 \times 0.0143 = 0.0714$.

The same logic can be applied to populations of different sizes, with the result in table 9.1.

TABLE 9.1 The Probability of Capturing a Single Marked Deer as a Function of Population Size

Population Size	Probability
8	0.0714
10	0.2381
15	0.4396
20	0.4696
25	0.4506
30	0.4196

The probability of catching exactly one marked deer is low for a small population because there is a high probability of catching more than one marked deer. As the population size increases, the probability of catching exactly one marked deer initially increases. Above a certain population size (here between twenty and twenty five) the probability starts to decrease because it becomes increasingly unlikely that we will catch any marked deer at all.

5. The probability of picking a good screw on your first try is 90/100. Of the remaining 99 screws, 89 are good, so the probability that your second screw is good is 89/99. For the third screw, the probability is 88/98, and so on. From the product rule (rule 6) we then find that the overall probability of picking ten good screws in a row is $90/100 \cdot 89/99 \cdot \ldots \cdot 81/91 = 0.3305$.

The probability of picking a bad screw on your first try, followed by picking nine good screws in a row, is $10/100 \cdot 90/99 \cdot 89/98 \cdot \ldots \cdot 82/91 = 0.0408$. The same probability is obtained if the sole bad screw is chosen on the second try, or third, or fourth, etc. The overall probability of choosing exactly one bad screw in ten tries is thus $0.0408 \cdot 10 = 0.4080$.

6. As suggested in the chapter, we work with the probability of *not* having any two people share the same birthday. We randomly choose a person to enter the room. With only one person in the room, it is absolutely certain that two people in the room don't have the same birthday. A second person is then chosen at random to enter the room. There are 364 ways in which the birthday of this second person can be different from that of the first, so the probability that the second person's birthday is different is 364/365. A third person is then chosen at random to enter the room, and there are 363 ways in which this person's birthday can differ from everyone else's in the room. The probability of a different birthday for this third person is thus 363/365. Because all three people were chosen at random, each choice is an independent event. As a result, we can use the product rule (rule 6) to calculate the probability that all three people have different birthdays: $1 \cdot 364/365 \cdot 363/365 = 0.9918$. The probability that among three people chosen at random at least two have the *same* birthday is therefore $1 - 0.9918 = 0.0082$ (see rule 2). We can extend this logic as more people are chosen to enter the room. With a bit of diligent bookkeeping, we find that with twenty three people in the room, there is a 50.73% chance that at least two of them share a birthday.

7. Switch. We were initially convinced that it made no difference whether you switched doors or not. Only when we wrote a computer program to simulate the game did the logic emerge. Try thinking of it this way. There are three doors but only one vacation, so the probability of picking the vacation on your first pick is 1/3. If you are lucky and make the correct choice in this first pick, you would be crazy to switch, because switching would mean that you inevitably lose. But let's examine the situation that arises if you make the *wrong* choice on your first pick. In this case, the vacation is behind one of the two remaining doors. The host (who knows what is where) opens the door with the cheese, leaving the vacation behind the *sole remaining door*, the door to which you would switch. In other words, if you are wrong in your first choice, switching doors will inevitably win you the vacation. Because your first choice is wrong 2/3 of the time, if you switch you win the vacation 2/3 of the time.

8. If there are four doors, you should still switch. Only one of the four doors hides the vacation, therefore you have only a one in four chance of picking it on the first try. In other words, if you do not switch, your probability of success is 1/4. What happens if you are wrong on your first guess and subsequently switch? The host opens one of the remaining three doors, revealing a cheese, and you then have to choose between two doors, only one of which hides the vacation. Therefore, if you were initially wrong and then switch, the probability of getting the vacation is 1/2. Since your chance of guessing correctly on your first door choice is 1/4, the chance of guessing incorrectly is $1 - 1/4 = 3/4$. Therefore, your chance of getting the prize given that you switch is $3/4 \times 1/2 = 3/8$. Yes, your chance of winning by switching is less that 50%, but since $3/8 > 1/4$, you should switch anyway.

In general, if there are n doors but only one vacation, the chance of winning if you don't switch is

$$P(\text{don't switch}) = \frac{1}{n}, \tag{9.2}$$

and the chance of winning if you do switch is

$$P(\text{switch}) = \left(1 - \frac{1}{n}\right) \times \left(\frac{1}{n-2}\right). \tag{9.3}$$

The ratio of $P(\text{switch})/P(\text{don't switch})$ is

$$\frac{P(\text{switch})}{P(\text{don't switch})} = \frac{1/(n-2) - n/(n(n-2))}{1/n}$$

$$= \frac{n-1}{n-2}. \tag{9.4}$$

Because $(n-1)$ is always greater than $(n-2)$, you should always switch. The value of switching, however, decreases as the number of doors increases.

9. When you first tell the rumor to someone else ($k = 1$), it is not being returned to you. So $P(k = 1)$, the probability that at $k = 1$ the rumor is returned, is 0. At the next step, the initial recipient of the rumor picks someone to tell, choosing from the n other people in town. Only one of these n people is you, so the probability of you being picked is $1/n$, and the probability of you not being picked is $1 - (1/n)$. Thus,

$$P(k = 2) = 1 - (1/n). \tag{9.5}$$

The same probability applies at the next iteration, so the probability that the rumor has not returned to you after three repetitions is (by the product rule, rule 6)

$$P(k = 3) = \left[1 - (1/n)\right] \times \left[1 - (1/n)\right]$$
$$= \left[1 - (1/n)\right]^2 \tag{9.6}$$

In general,

$$P(k) = \left[1 - (1/n)\right]^{k-1}. \tag{9.7}$$

Much the same logic can be used to calculate the probability that at each iteration the rumor is told to someone who has not heard it before. For $k = 2$, only you and the initial recipient have heard the rumor, so $P(2) = 1 - (1/n)$ as before. For $k = 3$, there are two possibilities that the rumor could be told to someone who has already heard it (you and the initial recipient), so $P(3) = \left[1 - (1/n)\right] \times \left[1 - (2/n)\right]$. In general,

$$P(k) = \left[1 - (1/n)\right] \times \left[1 - (2/n)\right] \times \left[1 - (3/n)\right] \times \ldots \times \left[1 - \frac{k-1}{n}\right]. \tag{9.8}$$

When N people are told the rumor at each iteration, the same logic applies. The probability that after k iterations the rumor has not returned to you is

$$P(k) = \left[1 - (N/n)\right]^{k-1}. \tag{9.9}$$

The probability that the rumor has not been repeated to anyone is

$$P(k) = \left[1 - (N/n)\right] \times \left[1 - (2N/n)\right] \times \left[1 - (3N/n)\right]$$
$$\times \ldots \times \left[1 - \frac{(k-1)N}{n}\right]. \tag{9.10}$$

10. a. There are two potential fathers, only one of which is the thoroughbred. So, lacking any other information, we assume that there is a 1 in 2 chance of the thoroughbred being the father.

b. Call the event that the thoroughbred is the father, T, and the event that the colt has the genetic marker, G_{colt}. We are seeking the conditional probability,

$P(T \mid G_{colt})$. However, the information actually at hand relates to two other conditional probabilities. Because the thoroughbred has the genetic marker and all of his male offspring must get the gene, we know that $P(G_{colt} \mid T) = 1$. Since the genetics of the other stallion are unknown, the probability that the colt would have the genetic marker given the other stallion as the father (i.e., $P(G_{colt} \mid Stallion) = P(G_{colt} \mid T^c)$) depends on the probability that the other stallion has the marker, which we are told is 0.02. This looks like a case for Bayes' formula. Plugging the appropriate events into Bayes' formula (eq. 2.9) yields

$$P(T \mid G_{colt}) = \frac{P(G_{colt} \mid T) \times P(T)}{\left[P(G_{colt} \mid T) \times P(T)\right] + \left[P(G_{colt} \mid T^c) \times P(T^c)\right]}. \quad (9.11)$$

The only things we are missing are the prior probabilities, $P(T)$ (the probability that the thoroughbred is the father) and $P(T^c)$ (the probability that the other stallion is the father). Neither $P(T)$ nor $P(T^c)$ are conditioned by the genetic information. Our best estimates of these probabilities come from part (a), $P(T) = 0.5$ and $P(T^c) = 0.5$. So,

$$P(T \mid G_{colt}) = \frac{1 \times 0.5}{(1 \times 0.5) + (0.02 \times 0.5)} = 0.98. \quad (9.12)$$

It's payday.

c. If the mare has had the opportunity to play the extended field, the only thing that has changed is that there is a much bigger pool of potential fathers. This means our best estimate of the prior probability $P(T)$ (the probability that the thoroughbred is the father) drops substantially. Because $P(T^c) = 1 - P(T)$, it also means that $P(T^c)$ rises substantially. If we didn't know anything about the genetics of the colt, we would assume that the thoroughbred would now have only a one in thousand chance of being the father. Let's try Bayes' formula again, with $P(T) = 0.001$ and $P(T^c) = 0.999$:

$$P(T \mid G_{colt}) = \frac{1 \times 0.001}{[1 \times 0.001] + [0.02 \times 0.999]} = 0.048. \quad (9.13)$$

Despite the fact that the thoroughbred and colt share a rare marker, the likelihood they are related is now quite small.

d. This problem is very similar to situations encountered in a variety of criminal trials. Whether the issue is paternity or some crime, if the suspect is identified by a rare genetic marker that matches the evidence, the probability of guilt assigned by the genetic evidence depends critically on our prior estimate of the probability of the suspect's guilt. If you assume the suspect is one of only two possible perpetrators and therefore has a 50:50 chance of being guilty, the genetic evidence is damning. If prior to obtaining the genetic evidence you assume that the suspect is one of many, you could reach a very different conclusion. The range of results depends greatly on the rarity of the genetic marker. In this case the difference associated with the two prior probabilities is large because the marker occurs in 2% of random individuals. If the probability of randomly encountering the marker were much smaller (as it would be with a variety of genetic analyses used by forensic scientists), the conclusion would not be so sensitive to the prior assumption of guilt. Nonetheless, the probabilities that are calculated from genetic evidence inevitably include an estimate of the likelihood of guilt. This opens the door for lots of mischief buried within the details of the mathematical analysis (e.g., see Sullivan 1990).

9.2 *Chapter 3*

1. a. The expectation of this repeated Bernoulli process is (from eq. 3.51)

$$\mu_X = n\big[(p \cdot V) + (q \cdot W)\big]$$
$$= 10\left[\left(\frac{1}{6} \cdot \$50\right) + \left(\frac{5}{6} \cdot -\$10\right)\right] = 0. \tag{9.14}$$

You expect to wind up even.

b. The standard deviation of this repeated Bernoulli process is (from eq. 3.61)

$$\sigma_X = \sqrt{npq}|V - W|$$
$$= \sqrt{10 \cdot \frac{1}{6} \cdot \frac{5}{6}}|\$50 - (-\$10)| = \$70.71. \tag{9.15}$$

You can expect a lot of fluctuation in your account in the course of ten rolls of the die.

c. If you write out the sample space for the ten rolls, you find that only two situations result in losses in excess of your initial pot of $30: if you lose every time, your casino account is −$70; if you lose all but once, your account is −$40. Using the equation for the binomial distribution (eq. 3.49), we find that the probability of zero wins is 0.1615, and that of one win is 0.3230. The probability of getting *either* zero or one win (and therefore being broke at the end of ten rolls of the die) is the sum of these two probabilities, 0.4845. In other words, there is a 48.45% chance of being in debt to the casino after ten rolls. Even though you should expect to come out even, there is an uncomfortably large chance that you will lose all your money.

d. There are four scenarios (= outcomes) in which your initial $30 balance could dip to zero, at which point the casino cuts you off:

- You lose on your first three rolls.
- You lose on your first two rolls, win once, and then lose six times in a row.
- You lose on your first roll, win once, and then lose seven times in a row.
- You win on your first roll and then lose eight times in a row.

If you win twice in your first three rolls, there is no way you can lose enough money in the remaining rolls to go into debt. The probability of losing on your first three rolls is $(5/6)^3 = 0.5787$. The probability of the remaining three scenarios is each $(1/6) \cdot (5/6)^8 = 0.0388$. The overall probability of going broke through one of these scenarios is the sum of these four probabilities: 0.6950. This probability exceeds the 0.4845 calculated using the binomial distribution (which, in essence, assumes that the casino allows debts). Thus, when you are not allowed to run a temporary deficit, the probability of going broke increases.

2. There are many biological situations in which an organism is not allowed to "run a tab." An animal cannot starve to death, only to be reincarnated when its luck changes. In such cases, care must be taken when calculating probabilities using the binomial distribution.

3. It is most probable that your first win will be on the first try (see fig. 3.3). The expectation, however, is affected by those rare events in which it takes you a very large number of trials to achieve a single success. As a result, the mean is much larger than the *mode* (the most probable value). This is a prime example of how technical terminology can be misleading. In everyday usage, the terms "most probable" and "expected" are interchangeable. Not so in statistics.

4. There are n ways to pick the first booby, and (with only one bird in hand) there is no possibility of having a mating pair. There are $n - 1$ ways to pick a second bird, but (because one of these $n - 1$ uncaptured birds is the mate of the one you have already captured) there are only $n - 2$ ways of picking that do not result in you having a mating pair. There are $n - 2$ birds to choose from for your third pick, but two of these are mates of birds already in captivity. As a result, there are only $n - 4$ ways of picking the third bird that do not result in your having a mating pair. In general, when picking the kth bird, there are $n - 2k + 2$ choices that do not result in your having a mating pair. Multiplying the number of choices available on the first pick by the number available on the second pick and so on until you get to the kth pick gives the total number of ways in which you can pick k birds without capturing a mating pair (the product rule, rule 6 from chapter 2). In the formal notation of mathematics, this serial multiplication is symbolized as:

$$\prod_{i=1}^{k}(n - 2i + 2). \tag{9.16}$$

The total number of ways in which you can pick k birds from a population of n birds is

$$n \times (n - 1) \times (n - 2) \times \ldots \times (n - k + 1), \tag{9.17}$$

calculated using eq. 3.45:

$$\frac{n!}{(n - k)!} = \binom{n}{k}k!. \tag{9.18}$$

Expressing the number of ways you can capture birds without obtaining a mating pair (eq. 9.16) as a fraction of the total number of ways of capturing birds yields our final answer:

$$P(\text{no mating pair}) = \frac{\prod_{i=1}^{k}(n - 2i + 2)}{\binom{n}{k}k!}. \tag{9.19}$$

For example, if $n = 20$ and $k = 8$, the probability of not having a mating pair is about 9%.

5. Let's review the facts. Let the number of soldiers be S. If you test all the soldiers individually, you have to do S blood tests. Alternatively, we could test pooled groups of soldiers, and let's assume that there are n soldiers in each group. This means we will have $N = S/n$ groups. For simplicity, let's assume that N is an integer value. If any soldier in a group has the disease, you end up doing $n + 1$ blood tests (1 for the group and n for the soldiers). If no one in the group has the disease, you only do one test. For the moment, let's focus on a random group. If the expected number of tests for the group exceeds n (as it would if on average even one soldier has the disease), then we would have been better off just testing all the soldiers individually.

Let h be the probability that an individual soldier does *not* have the disease. For an entire group to be free of the disease, all members must be disease-free. This has probability h^n as long as the individuals in a group are independent samples (i.e., they don't live in the same barracks and therefore are unlikely to affect the disease status of one another). The chance that at least one member of the group will have the disease is thus $1 - h^n$. We therefore have two possible outcomes for each group: (1) no one in the group is infected (with a cost of one test and a probability of h^n), and (2) at least one soldier is infected (with a cost of $n + 1$ tests and a probability of $1 - h^n$). We can now calculate the expected number of blood tests (T) for the group:

$$E(T) = [1 \times h^n] + [(n + 1) \times (1 - h^n)]$$
$$= n(1 - h^n) + 1 \tag{9.20}$$

The question then boils down to when this expected number of tests is less than n, the number of tests that would be done without grouping. That is, when will

$$n(1 - h^n) + 1 < n? \tag{9.21}$$

This simplifies to

$$1 - nh^n < 0,$$
$$nh^n > 1, \tag{9.22}$$

and given that n and h are both positive,

$$h > \left(\frac{1}{n}\right)^{1/n}. \tag{9.23}$$

If this condition is met, grouping is less expensive than testing all the soldiers individually. For groups of one hundred, greater than 95.5% of soldiers must be healthy for group testing to be cost effective. For groups of one thousand, greater than 99.3% must be healthy for grouping to pay off.

Now $n = S/N$, so

$$h > \left(\frac{N}{S}\right)^{N/S}. \tag{9.24}$$

This function is shown in figure 9.1 for a variety of S's. For any given S, if h falls above the line, grouped testing is recommended, for h below the line, individual testing is cheaper. For a large army, h must be very high before group testing is cost effective.

Note that figure 9.1 is a bit biased in its representation of the answer. In the figure, it appears that as N increases, there is a monotonic decrease in the h required to make group testing practical. If the abscissa of the graph extended to higher values, you would in fact see that this trend is reversed. As N approaches S, the required h approaches 1. But the whole point of testing groups is to reduce the number of samples, so in practice it would make little sense for N to be nearly as large as S.

6. The trick to solving this problem is to break it down into a number of events. An event occurs when you get a card with a new player. The first event

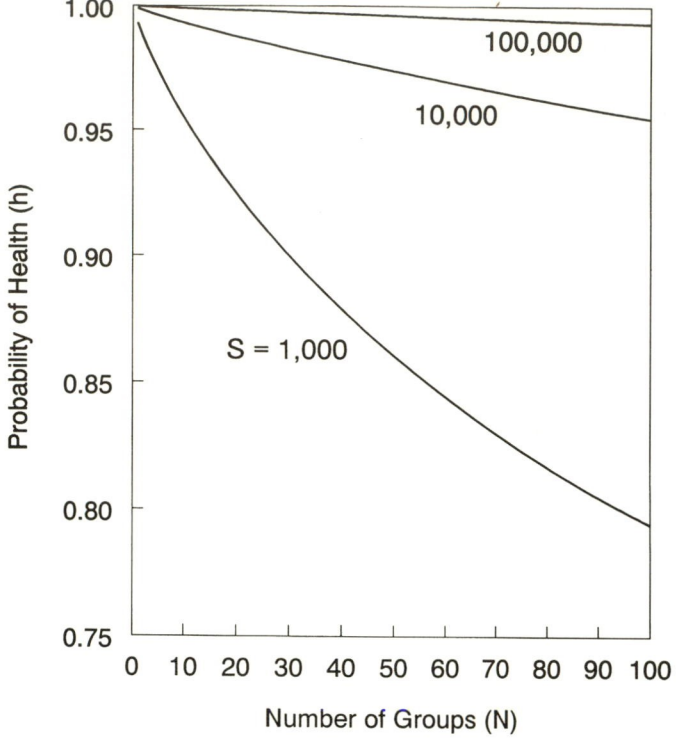

FIG. 9.1 The critical probability of health is shown as function of the number of groups tested (eq. 9.24) for armies with a total of 1000, 10,000, or 100,000 soldiers. For a given number of groups and size army, if h is greater than the value on the curve, grouped testing is cost effective. If h is less than the value on the curve, it is cheaper to test each soldier individually.

is easy. The first card you get will be a new player. Chance doesn't affect this outcome. After you have one card, however, getting a second player is a random event with uncertain outcome. In this case, since there are five different cards, after you have one card the chance that a new card will meet the criterion of being a new player is 4/5. So, how long can you expect to wait for a success if the probability of success is 4/5? This is the same question we addressed with the fringehead waiting for its first success. It concerns the probability of a string of failures followed by a single success (in this case success is getting a card with a player you do not have). The expected value of this geometric random variable (i.e., the average number of trips required to get a success) is $1/p$, where p is the probability of a success. To get all five cards, we need five successes where the respective probabilities are 1, 4/5, 3/5, 2/5, and 1/5.

Therefore, the total number of trips is the sum of the expected number of trips for each success:

$$= \frac{1}{1} + \frac{1}{4/5} + \frac{1}{3/5} + \frac{1}{2/5} + \frac{1}{1/5}$$

$$= 1 + \frac{5}{4} + \frac{5}{3} + \frac{5}{2} + 5$$

$$= 11.42 \text{ trips to the restaurant.} \qquad (9.25)$$

b. We use the same logic to calculate the mean number of trips required to get nine players. The probabilities of success are 1, 8/9, 7/9, 6/9, 5/9, 4/9, 3/9, 2/9, and 1/9. Therefore, the expected total number of trips is

$$= 1 + \frac{9}{8} + \frac{9}{7} + \frac{9}{6} + \frac{9}{5} + \frac{9}{4} + \frac{9}{3} + \frac{9}{2} + 9$$

$$= 25.46 \text{ trips to the restaurant.} \qquad (9.26)$$

How could you go about checking these answers? You could convince a number of your friends to each make a large number of trips to the restaurant, collecting cards as they go, but there is an easier way. This question—and many other similar ones in probability—is easily simulated on a computer. The following program can serve as a recipe for testing the answer to problem 6a. It is written in MATLAB but can easily be translated into your language of choice.

The program repeatedly calculates a random number, y, between 0 and 5. If $0 \leq y < 1$, the program assumes that it has received card 1, and if this is the first time it has gotten card 1 it changes the variable a from 0 to 1. If $1 \leq y < 2$, the program assumes that it has received card 2, and if this is the first time it has received card 2, it changes the variable b from 0 to 1. A similar procedure is followed for cards 3–5 and variables c–e. After each "card" is received, the program checks to see if it has at least one of each card. That is, it checks to see if $a + b + c + d + e = 5$. If this condition is met, it notes n, the number of cards it received in the process of getting at least one of each, and then repeats the whole process. The average value of n is our answer. When we ran this simulation for 10,000 trials, the average number of trips required to get all five cards was 11.48, quite close to the theoretical value of 11.42.

```
function cards
% initialize the random number generator
rand('state',sum(100*clock))
m = input('Number of trials?\n')
total = 0;
for i = 1:m
    a = 0;
    b = 0;
    c = 0;
    d = 0;
    e = 0;
    n = 0;
    y = 0;
    while y < 5
        n = n + 1;
        x = rand*5;
        if (x >= 0) & (x < 1)
            if a == 0; a = 1; end
        end
        if (x >= 1) & (x < 2)
            if b == 0; b = 1; end
        end
        if (x >= 2) & (x < 3)
            if c == 0; c = 1; end
        end
        if (x >= 3) & (x < 4)
            if d == 0; d = 1; end
        end
        if (x >= 4) & (x < 5)
            if e == 0; e = 1; end
        end
        y = a+b+c+d+e;
    end
    total = total+n;
end
mean = total/m;
fprintf(1, 'Mean = %2.4f\n', mean)
```

9.3 *Chapter 4*

1. a. The probability that the meteorite will strike in the range of longitudes that encompass Wyoming is simply the magnitude of that range ($7°$) divided by the magnitude of the entire range of longitudes (360°):

$$P(\text{longitude}) = \frac{7}{360} = 0.0194.$$ (9.27)

To calculate the probability that the meteorite will impact between any two given latitudes, imagine looking down on a photograph of the Earth taken from a position in space well above the North Pole. On this photograph, the lines of latitude can be drawn as concentric circles (as on a target). The radius of each circle of latitude ϕ is $r\cos\phi$, where r is the radius of the Earth. For each circle of latitude, the map area between it and the North Pole is

$$\text{enclosed area} = \pi(r\cos\phi)^2,$$ (9.28)

and the area between latitudes ϕ_1 and ϕ_2 is

$$A = \pi r^2\left(\cos^2\phi_1 - \cos^2\phi_2\right) \quad \phi_1 \le \phi_2.$$ (9.29)

The probability of impact between these two latitudes is equal to area A expressed as a fraction of the total map area, πr^2. Thus,

$$P(\text{latitude}) = \frac{\pi r^2\left(\cos^2\phi_1 - \cos^2\phi_2\right)}{\pi r^2}$$
$$= \cos^2 41° - \cos^2 45°$$
$$= 0.0696.$$ (9.30)

Because the probability of impacting at a given longitude is independent of the probability of impacting at a given latitude, the overall probability of impacting in Wyoming is the product of $P(\text{longitude})$ and $P(\text{latitude})$; $0.0194 \times 0.0696 = 0.0014$.

There are seven days in a week. The probability that impact will occur on any particular day (Tuesday) is thus $1/7 = 0.1429$. The probability that the meteorite hits Wyoming on a Tuesday is $0.0014 \times 0.1429 = 1.93 \times 10^{-4}$.

b. If the meteorite can approach from any direction, hitting the Earth entirely at random, the probability that it hits Wyoming is simply equal to the fraction of the total surface area of the Earth enclosed within Wyoming's boundaries. This area is calculated as follows.

At any latitude ϕ, the east-west distance across Wyoming is

$$w = 2\pi r \cos\phi \cdot \frac{7}{360}.$$ (9.31)

An infinitesimal north-south distance along the Earth's surface is $rd\phi$. The area of Wyoming is the integral of the product of east-west distance and north-south distance:

$$\text{surface area} = \int_{41}^{45} wr d\phi$$

$$= \frac{14\pi r^2}{360} \int_{41}^{45} \cos\phi d\phi$$

$$= \frac{14\pi r^2}{360} \cdot \sin\phi \mid_{41}^{45}$$

$$= 0.002\pi r^2. \tag{9.32}$$

Expressing this area as a fraction of the overall surface area of the Earth ($4\pi r^2$), we find that the probability of hitting Wyoming is 0.0005. Dividing by 7 to calculate the probability that impact occurs on a Tuesday, we arrive at our final answer. If the meteorite approaches from a random direction, the probability of hitting Wyoming on a Tuesday is 7.1×10^{-5}.

2. The probability density function is the first derivative of the cumulative probability distribution:

$$g(X) = \frac{d(x/10)^2}{dx} = \frac{x}{50}. \tag{9.33}$$

Probability density increases linearly through time.

The likelihood of dying between the ages of 3 and 4 years can be calculated either from the probability density function,

$$P(3 \le x \le 4) = \int_3^4 g(x) dx$$

$$= \frac{x^2}{100} \mid_3^4 = 0.07, \tag{9.34}$$

or from the cumulative probability distribution,

$$P(3 \le x \le 4) = P(x \le 4) - P(x \le 3)$$

$$= \frac{16}{100} - \frac{9}{100} = 0.07. \tag{9.35}$$

3. Transform the measured heights to Z values by first subtracting the mean (2.3 m) and then dividing by the standard deviation (0.5 m). The probability that a plant chosen at random has a height between 2.6 and 2.9 meters is thus the probability that Z has a value between 0.6 and 1.2. Consulting table 4.1 for the cumulative probability distribution of Z, we find that $P(Z \le 1.2) = 0.8849$.

$P(Z \leq 0.6) = 0.7257$. The difference in these two probabilities is the $P(0.6 \leq Z \leq 1.2) = 0.1592$, and this is the answer to our question.

Because there is a continuous distribution of heights, the probability that a plant is exactly 1.9 meters high is exactly zero.

4. Repeated throws of the die are a binomial process. If we define throwing a "3" as a success, $p = 1/6$ and $q = 5/6$. The expected (mean) number of successes in $n = 1000$ throws is $\mu_X = np \cong 167$ (eq. 3.51 with $V = 1$ and $W = 0$). The standard deviation around this mean is $\sqrt{npq} \cong 11.8$ (eq. 3.61). To calculate the probability that a "3" is thrown less than $x = 145$ times, we calculate

$$Z = \frac{x - \mu_X}{\sigma_X}$$

$$= \frac{145 - 167}{11.8}$$

$$= -1.86. \tag{9.36}$$

$P(-1.86) \cong 0.032$ (by interpolation from table 4.1). There is thus a probability of about 3.2% that in 1000 throws of a die, fewer than 145 "3's" will be obtained.

5. The standard deviation of the sample means (the standard error) is

$$\sigma_{\overline{X}} = \frac{\sigma_X}{\sqrt{n}}$$

$$= \frac{3}{\sqrt{10}} = 0.949. \tag{9.37}$$

The probability that the mean of a sample is greater than 15 mm is estimated using the standard normal curve. Given that $\mu_X = 14$,

$$Z = \frac{x - \mu_X}{\sigma_{\overline{X}}}$$

$$= \frac{15 - 14}{0.949} \cong 1.05. \tag{9.38}$$

From table 4.1 we find that the probability that $Z < 1.05$ is 0.8528. Therefore, the probability that the mean of a sample of ten fish is greater than 15 mm is $1 - 0.8528 = 0.1472$.

If we include hundred fish in our sample (rather than ten), the standard error is $(3/\sqrt{100}) = 0.3$. Z in this case is 3.33, and the probability that the mean of the sample is greater than 15 mm is less than 0.001. Increasing the sample size increases the reliability of the sample mean as an estimate of the population mean.

9.4 *Chapter 5*

1. The probability distribution of particle locations is calculated using eq. (3.49) for the binomial distribution. The number of successes is translated into a location on the *x*-axis by noting that a success results in moving one step to the right, while a loss results in moving one step to the left. Thus, zero successes implies 10 steps to the left and none to the right, for a final location at −10 steps. One success implies 9 steps to the left and one to the right for a final location of −8 steps, etc. The results are shown in figure 9.2.

When $p = 0.5$, the distribution is symmetrical about the origin, with $\mu_X = 0$ steps. When $p = 0.9$, the distribution is skewed to the right and $\mu_X = 8$ steps. When $p = 0.3$, the distribution is skewed to the left and $\mu_X = -4$ steps.

2. In this case, "speed" consists of both a steady drift in the mean location of particles and the diffusive spread about this mean location. First, the mean. From eq. (3.51) we know that

$$\mu_X = n\big[(p \cdot V) + (q \cdot W)\big]. \tag{9.39}$$

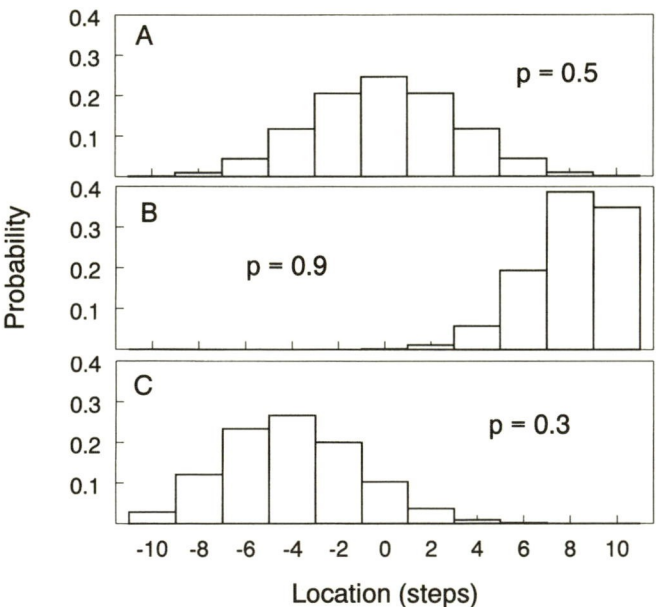

FIG. 9.2 The probability distribution of particle location for three values of p, the probability of stepping to the right.

In this case,

$$\mu_X = n\big[(0.6 \times 1 \text{ mm}) + (0.4 \times -1 \text{ mm})\big]. \tag{9.40}$$

Note that n (the number of steps) is equal to time divided by the interval between steps: $n = t/\tau$. Thus,

$$\mu_X = \frac{t \times \big[(0.6 \times 1 \text{ mm}) + (0.4 \times -1 \text{ mm})\big]}{\tau}. \tag{9.41}$$

In this problem, $\tau = 1$ s. Speed is distance per time, so the "drift speed" of particles due to the shift in the mean is

$$\text{drift speed} = \frac{\mu_X}{t} = \frac{\big[(0.6 \times 1 \text{ mm}) + (0.4 \times -1 \text{ mm})\big]}{1 \text{ s}} = 0.2 \text{ mm s}^{-1}. \tag{9.42}$$

The spread of particles about this mean is quantified by the standard deviation of particle location. From eq. (3.61) we know that

$$\sigma_X = \sqrt{npq}|V - W|. \tag{9.43}$$

In the terms of this problem:

$$\sigma_X = \sqrt{\frac{t \cdot 0.6 \cdot 0.4}{\tau}}|(1 \text{ mm}) - (-1 \text{ mm})|$$

$$= 0.9798 \cdot \sqrt{t}. \tag{9.44}$$

Again, we divide distance by time to calculate a speed:

$$\text{diffusive speed} = \frac{\sigma_X}{t} = \frac{0.9798}{\sqrt{t}}. \tag{9.45}$$

The overall speed is the sum of drift and diffusive speeds:

$$\text{speed} = 0.2 \pm \frac{0.9798}{\sqrt{t}} \text{ mm s}^{-1}. \tag{9.46}$$

In the right-hand tail of the particle distribution, the diffusive spread of particles to the right adds to the mean drift. In the left-hand tail of the distribution, the diffusive spread of particles to the left is subtracted from the mean drift of particles to the right.

3. From eq. (5.65) we know that the number of disks required to yield half the maximal current is

$$N_{1/2} = \frac{\pi r_s}{r_d}, \tag{9.47}$$

where r_d is the radius of each disk and r_s is the radius of the sphere. The area of each disk is πr_d^2, so the total area of disks is

$$N_{1/2}\pi r_d^2 = \frac{\pi r_s}{r_d} \cdot \pi r_d^2 = \pi^2 r_s r_d. \tag{9.48}$$

The surface area of the sphere is $4\pi r_s^2$. Thus, the fraction of the total area of the sphere that is occupied by disks at half maximum diffusive current is

$$\frac{\pi^2 r_s r_d}{4\pi r_s^2} = \frac{\pi r_d}{4r_s}. \tag{9.49}$$

The smaller the disks and the larger the sphere, the smaller the fraction of the sphere that needs to be occupied by disks to yield half the maximal current.

From eq. (5.64) we know that the overall current to disks on a sphere is

$$\text{overall diffusive current} = \frac{4\pi D r_s C_\infty}{1 + [\pi r_s/(N r_d)]}, \tag{9.50}$$

where N is the number of disks. Dividing by the number of disks, we find that the current per disk is

$$\text{current per disk} = \frac{4\pi D r_s C_\infty}{N + \pi r_s/r_d}. \tag{9.51}$$

The larger the number of disks, the smaller the current per disk.

These calculations tells us that when disks are tightly packed, each operates at less than maximum efficiency. As a result, we might expect evolution in real cells to have favored small receptors or ion channels evenly spaced on the cell surface. This would reduce the expenditure (in terms of receptors or channels produced) that is required to achieve a given diffusive current to the cell.

4. The speed at which molecules move is calculated by equating kinetic energy to thermal energy:

$$\frac{\overline{mu^2}}{2} = \frac{3kT}{2}, \tag{9.52}$$

where m is the mass of the molecule moving at speed u, k is Boltzmann's constant ($1.38 \cdot 10^{-23}$ joules per kelvin), and T is absolute temperature. Note that this is overall speed, rather than speed along any particular axis. Solving for the root mean square velocity:

$$\sqrt{\overline{u^2}} = \sqrt{\frac{3kT}{m}}. \tag{9.53}$$

Inserting values from the problem, we calculate an rms speed for oxygen molecules of 478 m s^{-1}. The mean free path is equal to the diffusion coefficient divided by this speed. In air, $L = 3.3 \cdot 10^{-8}$ m. This is roughly the diameter of the small tubes (tracheoles) that insects use to breath (see Denny 1993). In water, $L = 4.4 \cdot 10^{-12}$ m, a length small when compared even to individual atoms.

5. Because of its high swimming speed, the marine bacterium has an effective diffusion coefficient (4.5×10^{-9} m^2 s^{-1}) that is slightly larger than that of small molecules (about 1×10^{-9} m^2 s^{-1}) and an order of magnitude larger than that of the gut bacterium (4.1×10^{-10} m^2 s^{-1}).

From eq. (6.38) we know that $\overline{R^2} = n\delta^2$, where n is the number of steps taken in a three-dimensional random walk and δ is the length of each step. According to our rules of a random walk, $n = t/\tau$, where τ is the time between steps. Following the logic of eqs. (5.23) and (5.24), we calculate that

$$\overline{R^2} = \frac{t\delta^2}{\tau} = 2Dt. \tag{9.54}$$

Thus,

$$t = \frac{\overline{R^2}}{2D}. \tag{9.55}$$

Given that $\sqrt{\overline{R^2}} = 1 \times 10^{-3}$ m, we find that it takes an average of 111 seconds for the marine bacterium to diffuse 1 mm, 500 s for a small molecule, and 1,220 s for the gut bacterium.

It is tempting to suppose that the exceptional locomotory speeds of small marine bacteria (more than 1300 body lengths per second!) have evolved to allow these organisms better control of their position relative to the concentration of nutrient molecules in their environment.

9.5 *Chapter 6*

1. From eq. (6.19) we know that the overall average time to capture for particles confined between two absorbing barriers is

$$\frac{b^2}{12D},$$

where b is the distance between barriers (in this case, 5000 km) and D is the diffusion coefficient (in this case, 500 m^2 s^{-1}). Crunching through the math, we calculate an average time to capture of $4.17 \cdot 10^9$ s, approximately 132 years. Even with a very large diffusion coefficient, diffusive spread is an ineffective means of transport over large distances.

2. From eq. (6.45), we know that the probability density function (with respect to location x, y, z) for a particle undergoing a three dimensional random walk is

$$g(X, Y, Z) = \left(\frac{\sqrt{3}}{\sqrt{\overline{R^2}}\sqrt{2\pi}}\right)^3 \exp\left(-\frac{3r^2}{2\overline{R^2}}\right), \tag{9.56}$$

where r is the distance from the origin to the location of the particle at x, y, z. We also know that $\overline{R^2} = n\delta^2$ (eq. 6.38). Now, n (the number of steps) is equal to t/τ, where τ is the period between steps (in this case 1 s) and t is time. So, in this example we can replace $\overline{R^2}$ with $t\delta^2$ (where δ, the step length, is 5×10^{-5} m and t is measured in seconds):

$$g(X, Y, Z) = \left(\frac{\sqrt{3}}{\delta\sqrt{2\pi t}}\right)^3 \exp\left(-\frac{3r^2}{2t\delta^2}\right). \tag{9.57}$$

Now, $g(X, Y, Z)$ is the probability per volume that a particle is located at a particular location x, y, z, and to calculate the overall probability that a sperm lies within the volume of the egg we would need to integrate the probability density function across the egg volume. However, because the egg is small compared to the distance traveled by the sperm, we can approximate this integration by multiplying $g(X, Y, Z)$ (calculated for the location of the center of the egg) by the entire volume of the egg, $(4/3)\pi r_{egg}^3$. Thus,

$$P(r) \cong \frac{4}{3}\left(\frac{\sqrt{3}}{\delta\sqrt{2\pi t}}\right)^3 \exp\left(-\frac{3r^2}{2t\delta^2}\right)\pi r_{egg}^3. \tag{9.58}$$

This expression is graphed in fig. (9.3). The probability of fertilization reaches an abrupt peak at about 400 seconds after the sperm is released, and slowly decreases thereafter. For an extra bit of fun, vary the step length δ and see how it affects the location of the peak in the probability of fertilization.

3. From eq. (6.60) we know that the force F required to stretch a rubbery protein chain to an end-to-end length r is

$$F = kT\left(\frac{3r}{n\delta^2} - \frac{2}{r}\right),$$

where n is the number of amino acids in the protein chain, and δ is the length of each amino acid. Now, stiffness is change of force per change in extension (dF/dr), so

$$\text{stiffness} = kT\left(\frac{3}{n\delta^2} + \frac{2}{r^2}\right). \tag{9.59}$$

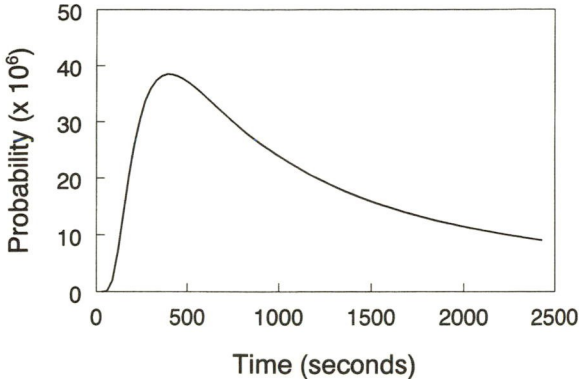

FIG. 9.3 The approximate probability of fertilization of the egg increases with time to a peak at about 400 seconds after sperm release and then gradually declines (eq. 9.58).

At large end-to-end separations, the term $2/r^2$ is small compared to $3/(n\delta^2)$, and we can neglect it. Thus, for the purposes of this question,

$$\text{stiffness} \cong \frac{3kT}{n\delta^2}. \tag{9.60}$$

Let n_a be the number of amino acids in the protein from the arctic jellyfish and n_t be the number in the tropical animal. Inserting the temperatures specified in the problem and equating the stiffness of the two rubbers, we see that

$$\frac{3k \cdot 273}{n_a \delta^2} = \frac{3k \cdot 303}{n_t \delta^2}. \tag{9.61}$$

Given that the length of amino acids is the same in the arctic as it is in the tropics,

$$\frac{n_t}{n_a} = \frac{303}{273} = 1.11. \tag{9.62}$$

The tropical jellyfish should have rubbery protein chains with about 11% more amino acids than the arctic jellyfish in order to keep the stiffness of the materials the same.

9.6 *Chapter 7*

1. The key to this problem is calculating the root mean square amplitude, a_{rms}. For each amplitude listed in the problem, calculate its square. Sum the squares for all 100 amplitudes, divide the sum by 100, and take its square root: $a_{rms} = 1.31$ m.

The probability that a randomly chosen amplitude is greater than 4 meters is (from eq. 7.6):

$$P(a \geq 4) = \exp\left(-\frac{4}{1.31}\right)^2 = 8.93 \cdot 10^{-5}. \tag{9.63}$$

The most probable maximum wave amplitude in a sample of thousand waves is (from Eq. 7.20)

$$\text{most probable } a_{max} = a_{rms}\sqrt{\ln(n)} = 1.31 \cdot \sqrt{\ln(1000)} = 3.44 \text{ m.} \tag{9.64}$$

The maximal wave amplitude expected in 24 hours is calculated using eq. (7.23):

$$\overline{a_{max}} = a_{rms}\left(\sqrt{\ln(ft)} + \frac{0.5772}{2\sqrt{\ln(ft)}}\right). \tag{9.65}$$

In this problem, the wave frequency $f = 1/11 = 0.091$ Hz, and there are roughly $3.15 \cdot 10^7$ s in a year. Thus,

$$a_{max} = a_{rms}\left(\sqrt{\ln(0.091 \cdot 3.15 \cdot 10^7)} + \frac{0.5772}{2\sqrt{\ln(0.091 \cdot 3.15 \cdot 10^7)}}\right)$$
$$= 3.93 \cdot a_{rms} = 5.15 \text{ m.} \tag{9.66}$$

2. The cumulative probability of these thirty values for wind speed is calculated as follows. First, rank the values in ascending order. Then assign each value a cumulative probability according to eq. (7.29):

$$P(x_{max} \leq x_i) = \frac{i}{(n+1)}.$$

Here i is a value's rank and $n = 30$. For example, the lowest wind speed (with rank 1) has cumulative probability

$$P(x_{max} \leq 0.741) = \frac{1}{31} = 0.032. \tag{9.67}$$

The resulting distribution is shown in figure 9.4.

The probability that the maximum wind speed for an interval chosen at random is less than or equal to 10 m s^{-1} is (from eq. 7.30):

$$P(x_{max} \leq 10) = \exp\left[-\frac{\alpha - (\beta \times 10)}{\alpha - \beta\epsilon}\right]^{1/\beta}$$
$$= \exp\left[-\frac{1.4306 - 0.05279}{1.4306 - (0.005279 \times 2.8307)}\right]^{1/0.005279}$$
$$= 0.99412. \tag{9.68}$$

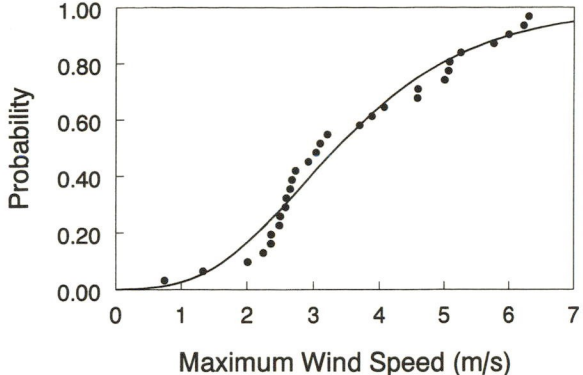

FIG. 9.4 The cumulative probability curve for maximal wind speeds. Note that the theoretical curve (the solid line) does not fit closely to the empirical data (the dots).

The associated return time is (from eq. 7.33):

$$\text{return time} = \frac{\text{interval length}}{1 - P(x_{max} \leq 10)}$$

$$= \frac{2 \text{ hours}}{1 - 0.99412}$$

$$= 340 \text{ hours.} \tag{9.69}$$

Your confidence in this prediction should be tempered by the fact that the asymptotic curve in figure 7.19 is not a very tight fit to the empirical data. Mind you, it is the *best* fit, but in this case best is not very good. For an in-depth discussion of how to calculate the confidence intervals for a prediction of return time, consult Gaines and Denny (1993).

The absolute maximum wind speed predicted by this analysis is $\alpha/\beta = 271 \text{ m s}^{-1}$.

9.7 *Chapter 8*

The energy associated with an infinitesimal deflection of the pendulum is the integral of force ($mg\theta$) times distance ($Ld\theta$). Integrating over distance,

$$\text{energy} = \int mg\theta Ld\theta$$

$$= mgL \int \theta d\theta$$

$$= \frac{mgL\theta^2}{2}. \tag{9.70}$$

Equating the average deflection energy with the unavoidable thermal energy, we see that

$$\frac{\overline{mgL\theta^2}}{2} = \frac{kT}{2}, \tag{9.71}$$

where k is Boltzmann's constant and T is absolute temperature. Assuming that mass m and the acceleration due to gravity g are constant, we solve for the root mean square deflection of the pendulum:

$$\theta_{rms} = \sqrt{\overline{\theta^2}} = \sqrt{\frac{kT}{mgL}}. \tag{9.72}$$

As the mass and length of the pendulum decrease, the noise (θ_{rms}) increases.

Averaging values of θ through time decreases the noise. But the fluctuations in this mean value depend on N, the number of samples that go into the average. The standard deviation of the mean (the standard error) is (from eq. 4.24)

$$\text{standard error} = \frac{\theta_{rms}}{\sqrt{N}}. \tag{9.73}$$

The number of samples is determined by t (the time over which you take data) and the period of the pendulum's swing:

$$N = \frac{t}{\text{period}} = \frac{t\sqrt{g}}{2\pi\sqrt{L}}. \tag{9.74}$$

Thus, the standard error (an index of the fluctuation in your mean values of θ) is

$$\text{standard error} = \frac{\sqrt{2\pi}\,\theta_{rms}L^{1/4}}{\sqrt{t}\,g^{1/4}}. \tag{9.75}$$

Inserting our expression from above for θ_{rms} and sorting things out, we find that

$$\text{standard error} = \frac{\sqrt{2\pi kT}}{t^{1/2}m^{1/2}L^{1/4}g^{3/4}}. \tag{9.76}$$

You can reduce the "noise" in your measurement by

- Decreasing the temperature.
- Increasing the time over which you take an average.
- Increasing the mass at the end of the pendulum.
- Increasing the pendulum's length.

Of these options, increasing the length is the least effective because L is present in the equation only as a fourth root, whereas time and mass are present as square roots.

Literature Cited

Abramowitz, M., and I. Stegun. 1965. *Handbook of Mathematical Functions*. Dover, New York.

Aklonis, J.J., W.J. MacKnight, and M. Shen. 1972. *Introduction to Polymer Viscoelasticity*. Wiley-Interscience, New York.

Alexander, R. McN. 1983. *Animal Mechanics*. 2nd ed. Blackwell Scientific, Boston.

Alexander, R. McN. 1992. *Exploring Biomechanics: Animals in Motion*. Scientific American Library, New York.

Ali, M.A., and M.A. Klyne. 1985. *Vision in Vertebrates*. Plenum Press, New York.

Atkins, P.W. 1984. *The Second Law*. Scientific American Library. New York.

Baylor, D. 1995. Colour mechanisms of the eye. In T. Lamb and J. Bourriau (eds.), *Colour: Art and Science*, 103–126. Cambridge University Press. Cambridge. U.K.

Baylor, D.A., T.D. Lamb, and K.W. Yau. 1979. Responses of retinal rods to single photons. *J. Physiol.* 288:613–634.

Bell, E.T. 1937. *Men of Mathematics*. Simon and Schuster, New York.

Benzi, R., G. Parisi, A. Sutera, and A. Vulpiani. 1982. Stochastic resonance in climatic change. *Tellus* 34:10–18.

Berg, H.C. 1983. *Random Walks in Biology*. Princeton University Press, Princeton, N.J.

Berg, H.C., and E.M. Purcell. 1977. Physics of chemoreception. *Biophys. J.* 20:193–219.

Bernstein, P.L. 1996. *Against the Gods: The Remarkable Story of Risk*. John Wiley and Sons, New York.

Bezrukov, S.M., and I. Vodyanoy. 1995. Noise induced enhancement of signal transduction across voltage-dependent ion channels. *Nature* 378:362–364.

Bezrukov, S.M., and I. Vodyanoy. 1997. Stochastic resonance in non-dynamical systems without response thresholds. *Nature* 385:319–321.

Bialek, W. 1987. Physical limits to sensation and perception. *Ann. Rev. Biophys. Chem.* 16:455–478.

Block, S.M. 1992. Biophysical principles of sensory transduction. In D.P. Corey and S.D. Roper (eds.), *Sensory Transduction*, 2–16. 45th Ann. Symposium, Society of General Physiologists. Rockefeller University Press, New York.

Bulsara, A.R., and L. Gammaitoni. 1996. Tuning in to noise. *Physics Today* 49(3):39–45.

Charlesworth, B. 1980. *Evolution in Age-Structured Populations*. Cambridge University Press, Cambridge, U.K.

Cohen, J.E. 1976. Irreproducible results and the breeding of pigs. BioScience 26:391–394.

Collins, J.J., C.C. Chow, and T.T. Imhoff. 1995. Stochastic resonance without tuning. *Nature* 376:236–238.

Cornsweet, T.N. 1970. *Visual Perception*. Academic Press, New York.

Csanady, G.T. 1973. *Turbulent Diffusion in the Environment*. D. Reidel, Boston.

Denny, M.W. 1980. Silks—their properties and functions. In J.F.V. Vincent and J.D. Currey (eds.), *The Mechanical Properties of Biological Materials*. Symp. Soc. Exp. Biol. xxxiv.

Denny, M.W. 1988. *Biology and the Mechanics of the Wave-Swept Environment.* Princeton University Press, Princeton, N.J.

Denny, M.W. 1993. *Air and Water.* Princeton University Press, Princeton, N.J.

Denny, M.W. 1995. Predicting physical disturbance: Mechanistic approaches to the study of survivorship on wave-swept shores. *Ecol. Monogr.* 65:371–418.

Donner, K. 1992. Noise and the absolute threshold of cone and rod vision. *Vision Res.* 32:852–866.

Douglass, J.K., L. Wilkens, E. Pantazelou, and F. Moss. 1993. Noise enhancement of information transfer in crayfish mechanoreceptors by stochastic resonance. *Nature* 365:337–340.

Dunthorn, D. 1996. Try, try, and try again. *Nature* 380:477.

Dykman, M.I., and P.V.E. McClintock. 1998. What can stochastic resonance do? *Nature* 391:344.

Einstein, A. 1905. On the motion of small particles suspended in liquids at rest required by the molecular-kinetic theory of heat. Translated and annotated in Stachel, J. (ed.), *Einstein's Miraculous Year: Five Papers That Changed the Face of Physics.* Princeton University Press, Princeton, N.J.

Feller, W. 1960. *An Introduction to Probability Theory and Its Application.* John Wiley and Sons, New York.

Gaines, S.D., and M.W. Denny. 1993. The largest, smallest, highest, lowest, longest, and shortest: Extremes in ecology. *Ecology* 74:1677–1692.

Gegenfurtner, K.E., H. Mayser, and L.T. Sharpe. 1999. Seeing movement in the dark. *Nature* 398:475–476.

Gonick, L., and W. Smith. 1993. *The Cartoon Guide to Statistics.* Harper Perennial, New York.

Gosline, J.M., M.W. Denny, and E.M. Demont. 1984. Spider silk as rubber. *Nature* 309:551–552.

Gosline, J.M., E.M. Demont, and M.W. Denny. 1986. The structure and properties of spider silk. *Endeavor* 10(1):37–44.

Gumbel, E.J. 1958. *Statistics of Extremes.* Columbia University Press, New York.

Hecht, S., S. Shlaer, and M.H. Pirenne. 1942. Energy, quanta, and vision. *J. Gen. Physiol.* 25:819–840.

Hille, B. 1992. *Ionic Channels of Excitable Membranes.* 2nd ed. Sinauer Associates, Sunderland, Mass.

Hoerner, S.F. 1965. *Fluid-Dynamic Drag.* Hoerner Press. Bricktown, N.J. (published by the author, now deceased).

Hoffman, P. 1998. *The Man Who Loved Only Numbers.* Hyperion Press, New York.

Inglis, J.T., S. Vershueren, J.J. Collins, D.M. Merfeld, S. Rosenblum, S. Buckley, and F. Moss. 1996. Noise in human muscle spindles. *Nature* 383:769–770.

Isaac, R. 1995. *The Pleasure of Probability.* Springer-Verlag, New York.

Jacocks, J.L., and K.R. Kneile. 1975. Statistical prediction of maximum time-variant inlet distortion levels. Arnold Engineering Development Center, Technical Report AD/A-004. National Technical Information Service, U.S. Department of Commerce.

Killion, M.C. 1978. Revised estimate of minimum audible pressure: Where is the missing 6 dB? *J. Acoust. Soc. Am.* 63:1510–1508.

Kimura, M. 1954. Solution of a process of random genetic drift with a continuous model. *Proc. Natl. Acad. Sci. (USA)* 41:144–150.

Kinney, J.J. 1997. *Probability: An Introduction with Statistical Applications.* John Wiley and Sons, New York.

Lehninger, A.L. 1970. *Biochemistry*. Worth, New York.

Levin, J.E., and J.P. Miller. 1996. Broadband neural encoding in the cricket cercal sensory system enhanced by stochastic resonance. *Nature* 380:165–168.

Longuet-Higgins, M.S. 1952. On the statistical distribution of the height of sea waves. *J. Mar. Res.* 11:245–266.

Love, R.M. 1991. *Probably More than You Want to Know about the Fishes of the Pacific Coast*. Really Big Press, Santa Barbara, Calif.

Lynch, J.W., and P.H. Barry. 1989. Action potentials intiated by single channels opening in a small neuron (rat olfactory receptor). *Biophys. J.* 55:755–768.

Magariyama, Y., S. Sugiyama, N. Muramoto, Y. Maekawa, I. Kawagishi, Y. Imae, and S. Kudo. 1994. Very fast flagellar rotation. *Nature* 371:752.

Mann, K.H., and J.R.N. Lazier. 1991. *Dynamics of Marine Ecosystems*. Blackwell Scientific, Boston.

Matthews, R.A.J. 1995. Tumbling toast, Murphy's law, and fundamental constants. *Eur. J. Phys.* 16:172–176.

Matthews, R.A.J. 1997. The science of Murphy's law. *Sci. Amer.* 276(4):88–91.

May, R.M. 1976. Irreproducible results. *Nature* 262:646.

McNair, J.N., J.D. Newbold, and D.D. Hart. 1997. Turbulent transport of suspended particles and dispersing benthic organisms: How long to hit bottom? *J. Theor. Biol.* 188:29–52.

Milankovitch, M. 1930. *Handbunch der Klimatologie*. Part I. A.W. Koppen and W. Wegener (eds.). Doppen and Geiger, Publishers.

Mitchell, J.G., L. Pearson, A. Bonazinga, S. Dillon, H. Khouri, and R. Paxinos. 1995a. Long lag times and high velocities in the motility of natural assemblages of marine bacteria. *Appl. Environ. Microbiol.* 61:877–882.

Mitchell, J.G., L. Pearson, S. Dillon, and K. Kantalis. 1995b. Natural assemblages of marine bacteria exhibiting high-speed motility and large accelerations. *Appl. Environ. Microbiol.* 61:4436–4440.

Moon, F.C. 1992. *Chaotic and Fractal Dynamics*. John Wiley and Sons, New York.

Moss, F., and X. Pei. 1995. Neurons in parallel. *Nature* 376:211–212.

Neter, J., W. Wasserman, and M.H. Kutner. 1996. *Applied Linear Regression Models*. 3d ed. University of Chicago Press, Chicago.

Nicolis, C. 1982. Stochastic aspects of climate transitions—response to a periodic forcing. *Tellus* 34:1–9.

Norman, G.R., and D.L. Steiner. 1993. *Biostatistics: The Bare Essentials*. Mosby Press, St. Louis.

Okubo, A. 1980. *Diffusion and Ecological Problems: Mathematical Models*. Springer-Verlag, New York.

Peterson, I. 1993. *Newton's Clock: Chaos in the Solar System*. W.H. Freeman, New York.

Purcell, E.M. 1977. Life at low Reynolds number. *Amer. J. Physics* 45:3–11.

Rayleigh, J.W.S. 1880. On the resultant of a large number of vibrations of the same pitch and arbitrary phase. *Phil. Mag.* 10:73–78.

Rayleigh, J.W.S. 1945. *The Theory of Sound*. 2 vols. Dover Publications. (Reprint of 2nd ed., published originally in 1894.)

Reif, F. 1965. *Fundamentals of Statistical and Thermal Physics*. McGraw-Hill, New York.

Rice, S.O. 1944, 1945. Mathematical analysis of random noise. (Excerpted from *Bell Systems Tech. J.* 23 and 24 in Wax, N., ed., *Selected Papers on Noise and Stochastic Processes*. Dover Publications, New York, 1954.)

Ross, S.M. 1997. *Introduction to Probability Models*. 6th ed. Academic Press, San Diego, Calif.

Rossing, T.D. 1990. *The Science of Sound*. 2nd ed. Addison-Wesley, New York.

Ruhla, C. 1992. *The Physics of Chance*. Oxford University Press, Oxford, U.K.

Russell, D.F., L.A. Wilkins, and F. Moss. 1999. Use of behavioral stochastic resonance by paddle fish or feeding, *Nature* 402:291–294.

Schnitzer, M.J., and S.M. Block, H.C. Berg, and E.M. Purcell. 1990. Strategies for chemotaxis. *Symp. Soc. Gen. Microbiol.* 46:15–34.

Schnapf, J.L., and D.A. Baylor. 1987. How photoreceptor cells respond to light. *Sci. Amer.* 256(4):40–47.

Shadwick, R.E., and J.M. Gosline. 1983. Molecular biomechanics of protein rubbers in molluscs. In P.W. Hochachka (ed.), *The Mollusca*, vol. 1, *Metabolic Biochemistry and Molecular Biomechanics*, 399–430. Academic Press, New York.

Shu, F.H. 1982. *The Physical Universe: An Introduction to Astronomy*. University Science Books, Mill Valley, Calif.

Smith, K.K., and W.M. Kier, 1989. Trunks, tongues and tentacles: Moving with skeletons of muscle. *Amer. Sci.* 77:28–35.

Sokal, R.R., and F.J. Rohlf. 1995. *Biometry*. 3d ed. W.H. Freeman, New York.

Stokes, D. 1998. Biting sarcasm. *California Wild* 51(2):42.

Stryer, L. 1987. The molecules of visual excitation. *Sci. Amer.* 255:42–50.

Sullivan, J.F. 1990. Paternity test at issue in New Jersey sex assault case. *New York Times*, November 28.

Thorn, J., P. Palmer, M. Gershman, and D. Pietrusza. 1997. *Total Baseball*. 5th ed. Viking Press, Penguin Books USA, New York.

Tierney, J. 1991. Behind Monty Hall's doors: Puzzle, debate, and answer? *New York Times*, July 21.

Timoshenko, S.P. 1953. *History of Strength of Materials*. McGraw-Hill, New York.

Treloar, L.R.G. 1975. *Physics of Rubber Elasticity*. 3d ed. Clarendon Press, London.

van der Velden, H.A. 1944. Over het aantal Lichtquanta dat nodig is voor een Lichtprikkel bij het Menselik oog. *Physica* 11:179–189.

Vogel, S. 1994. *Life in Moving Fluids*. 2nd ed. Princeton University Press. Princeton, N.J.

Wainwright, S.A., W.D. Biggs, J.D. Currey, and J.M. Gosline. 1976. *Mechanical Design in Organisms*. Princeton University Press, Princeton, N.J.

Wiesenfeld, K., and F. Moss. 1995. Stochastic resonance and the benefits of noise: From ice ages to crayfish and SQUIDS. *Nature* 373:33–36.

Williams, B. 1993. *Biostatistics: Concepts and Applications for Biologists*. Chapman and Hall, London.

Williams, G.C. 1957. Pleiotropy, natural selection, and the evolution of senescence. *Evolution* 11:398–411.

Wolf, R. (ed.) 1990. *The Baseball Encyclopedia*. Macmillan, New York.

Young, M.C. (ed.) 1997. *The Guinness Book of World Records*. Bantam Books, New York.

Zar, J.H. 1996. *Biostatistical Analysis*. 3d ed. Prentice Hall, Englewood Cliffs, N.J.

Symbol Index

The symbols a, b, c, d, k, and x serve as our general-purpose variables. We have taken care to define their meaning in each instance where they are used, but you should be aware that their usage varies throughout the text. Specialized usage of these variables is tabulated here. A few other symbols have been used for multiple purposes, and their various definitions are listed below. Symbols used as names of outcomes and events are not listed here. The Greek alphabet follows the Roman alphabet.

VARIABLES AND FUNCTIONS

Symbol	Definition	Page Where Defined
a	amplitude of a single wave	176, 178
a_{max}	an individual maximum wave amplitude	183
a_{rms}	root mean square wave amplitude	178
A	area	123
	wave amplitude	178
A_i	an index for the presence of an allele	149
A_{max}	maximum wave amplitude	183
b	location of barrier on the x-axis	140
B	Bernoulli random variable	49
C	constant	45
	concentration	124
C_d	drag coefficient	188
C_∞	bulk concentration	126
$\mathcal{C}$	total number of conformations	170
d	diameter	229
D	diffusion coefficient	115
D_g	genetic diffusion coefficient	150
D_r	rotational diffusion coefficient	138
$E(X)$	expectation of X	44
f	failure	18
	frequency	176
F	force	156
g	acceleration due to gravity	247
$g(X)$	probability density function of X	72
G	geometric random variable	61

ϕ	phase of a wave	176
	latitude	263
$\Omega(x)$	number of possible states of x	170

STANDARD FUNCTIONS AND OPERATORS

Symbol	Definition	Page Where First Used
$\dfrac{df}{dx}$	derivative of f with respect to x	72, 125
$\exp(x)$	e^x	81
$\ln(x)$	natural logarithm of x	94
$P(X)$	probability of X	19
$P(X|Y)$	probability of X given Y	27
X^c	complement of X	26
$\dbinom{n}{i}$	combinations of n things taken i at a time	53
Π	product operator	258
Σ	summation operator	21
$\cup$	union operator	22
$\cap$	intersection operator	24
$\int f(x)dx$	integral of $f(x)$	72
$!$	factorial operator	54

Author Index

Abramowitz, M., and I. Stegun (1965), 153
Aklonis, J.J., W.J. MacKnight, and M. Shen (1972), 163
Alexander, R. McN. (1983), 159
Alexander, R. McN. (1992), 158
Ali, M.A., and M.A. Klyne (1985), 218
Atkins, P.W. (1984), 108

Baylor, D. (1995), 212
Baylor, D.A., T.D. Lamb, and K.W. Yau (1979), 213, 219
Bell, E.T. (1937), 48, 156
Benzi, R., G. Parisi, A. Sutera, and A. Vulpiani (1982), 244
Berg, H.C. (1983), xiii, 127, 130, 137, 138, 139, 140, 173
Berg, H.C., and E.M. Purcell (1977), 122, 130, 137, 245
Bernstein, P.L. (1996), 82
Bezrukov, S.M., and I. Vodyanoy (1995), 244
Bezrukov, S.M., and I. Vodyanoy (1997), 244
Bialek, W. (1987), 239, 245
Block, S.M. (1992), 239
Bulsara, A.R., and L. Gammaitoni (1996), 242

Charlesworth, B. (1980), 195
Cohen, J.E. (1976), 154, 156
Collins, J.J., C.C. Chow, and T.T. Imhoff (1995), 244, 245
Cornsweet, T.N. (1970), 213
Csanady, G.T. (1973), 148, 173

Denny, M.W. (1980), 4
Denny, M.W. (1988), 187, 188, 189
Denny, M.W. (1993), 107, 120, 130, 137, 269
Denny, M.W. (1995), 189
Donner, K. (1992), 225
Douglass, J.K., L. Wilkens, E. Pantazelou, and F. Moss (1993), 244
Dunthorn, D. (1996), 156
Dykman, M.I. and P.V.E. McClintock (1998), 244, 245

Einstein, A. (1905), 210

Feller, W. (1960), 12, 38, 39

Gaines, S.D., and M.W. Denny (1993), 194, 196, 206, 273
Gegenfurtner, K.R., H. Mayser, and L.T. Sharpe (1999), 216
Gonick, L., and W. Smith (1993), 90
Gosline, J.M., M.W. Denny, and E.M. Demont (1984), 4, 5
Gosline, J.M., E.M. Demont, and M.W. Denny (1986), 3
Gumbel, E.J. (1958), 190, 192, 206

Hecht, S., S. Shlaer, and M.H. Pirenne (1942), 213
Hille, B. (1992), 220, 228
Hoerner, S.F. (1965), 188
Hoffman, P. (1998), 39

Inglis, J.T., S. Verschueren, J.J. Collins, D.M. Merfeld, S. Rosenblum, S. Buckley, and F. Moss (1996), 244
Isaac, R. (1995), 12, 30

Jacocks, J.L., and K.R. Kneile (1975), 206

Killion, M.C. (1978), 231
Kimura, M. (1954), 153
Kinney, J.J. (1997), 64

Lehninger, A.L. (1970), 169
Levin, J.E., and J.P. Miller (1996), 244
Longuet-Higgins, M.S. (1952), 185
Love, R.M. (1991), 14
Lynch, J.W., and P.H. Barry (1989), 229

Magariyama, Y., S. Sugiyama, N. Muramoto, Y. Maekawa, I. Kawagishi, Y. Imae, and S. Kudo (1994), 122
Mann, K.H., and J.R.N. Lazier (1991), 146, 147, 148, 173
Matthews, R.A.J. (1995), 41
Matthews, R.A.J. (1997), 41
May, R.M. (1976), 156
McNair, J.N., J.D. Newbold, and D.D. Hart (1997), 154

Subject Index